Hand Preference and Hand Ability

Advances in Interaction Studies (AIS)

Advances in Interaction Studies (AIS) provides a forum for researchers to present excellent scholarly work in a variety of disciplines relevant to the advancement of knowledge in the field of interaction studies. The book series accompanies the journal *Interaction Studies: Social Behaviour and Communication in Biological and Artificial Systems.*

The book series allows the presentation of research in the forms of monographs or edited collections of peer-reviewed material in English.

For an overview of all books published in this series, please see
http://benjamins.com/catalog/ais

Volume 5

Hand Preference and Hand Ability. Evidence from studies in Haptic Cognition
by Miriam Ittyerah

Hand Preference and Hand Ability

Evidence from studies in Haptic Cognition

Miriam Ittyerah
Institute of Communicative and Cognitive Neuroscience, India

John Benjamins Publishing Company
Amsterdam / Philadelphia

The paper used in this publication meets the minimum requirements of the American National Standard for Information Sciences – Permanence of Paper for Printed Library Materials, ANSI Z39.48-1984.

Library of Congress Cataloging-in-Publication Data

Ittyerah, Miriam.

Hand Preference and Hand Ability : Evidence from studies in Haptic Cognition / Miriam Ittyerah.

p. cm. (Advances in Interaction Studies, ISSN 1879-873X ; Volume 5)

Includes bibliographical references and index.

1. Touch. 2. Hand. 3. Left- and right-handedness. I. Title.

BF275.I88 2013

152.3'35--dc23 2013018483

ISBN 978 90 272 0459 2 (Hb ; alk. paper)

ISBN 978 90 272 7164 8 (Eb)

John Benjamins Publishing Co. · P.O. Box 36224 · 1020 ME Amsterdam · The Netherlands

John Benjamins North America · P.O. Box 27519 · Philadelphia PA 19118-0519 · USA

Table of contents

Preface

This book is product of my learning over the years since my doctoral work at the University of Delhi and subsequent post doctoral research at the Department of Experimental Psychology, University of Oxford. I had the good fortune of working with Susanna Millar at Oxford University during which time I learnt much about haptic processing. We demonstrated movement imagery in congenitally blind children in the total absence of vision and verbal mediation. I realized it was necessary to work with blind children or adults to know tactile processing, because the experience of sighted blindfolded subjects who participated as controls during experiments, were not devoid of visual experiences, particularly visual imagery.

An important aspect of haptic processing is the use of the hands for feeling the shape, size, contour or texture of an object. This in itself generated an interest in hand preference and the need to know if the preferred hand is the skilled hand. I considered it interesting to test the hand preference of blind children because most blind children read Braille with both hands. If Braille is read successfully with both hands, then the assumption that the preferred hand is the skilled hand may be questioned. My study in the subject of handedness in Neuro Psychology demonstrated hand brain relationships of tactile recognition of objects and proprioception for the first time in the total absence of vision. Studies with congenitally blind children over time revealed, that although one may have a preferred hand, the hand ability of the preferred and non preferred hands may not differ and therefore general laterality does not affect ability.

Interest in haptic processing continued as I worked with Lawrence E Marks at the John B Pierce Laboratory, Yale University. We designed experiments to show haptic processing mostly incorporated spatial and movement factors that are not affected by verbal mediation. Collaborations with Yves Rossetti and his team from INSERM, Lyon, France, at Delhi University investigated pointing with the left and right hands to immediate and delayed instructions in the total absence of vision. Findings revealed that differences between the hands were in the orientation they adopted during performance. The accuracy of the preferred hand was an outcome of its context related orientation, whereas the non preferred hand adopted an egocentric orientation, and was therefore less accurate. Even then, each hand demonstrated a potential ability to perform.

All this learning helped to conceive the idea of a book on the subject. During my visit at the Rockefeller Foundation at Bellagio, Italy, I wrote an outline for the proposed book. Although the ideas did not develop into a book immediately, I have been presently able to put forth most of it in separate chapters. Each chapter reveals how vast and unfathomable the sensory modality of touch and haptic process is likely to be. Besides, the role of hands is fundamental in haptics and no animal species can compare to humans in haptic performance. The present book is an interesting complement to *Human hand Function* by Lynette A. Jones and Susan J. Lederman. Though both books differ in scope, goals and approach, an attempt to combine touch, haptic processing, hand preference and hand ability in successive chapters is the prime concern of this book.

About the author

Miriam Ittyerah is Professor of Psychology. She has worked in the subject of handedness in Neuro Psychology with congenitally blind children and demonstrated hand brain relationships of tactile recognition of objects (haptics) and proprioception for the first time in the total absence of vision.

CHAPTER 1

Ways of knowing

1.1 Introduction

The impressions arising from sensations evoked by a wet cup or a dog's nose or slithering smoothness or dry flakes or viscous toffee, will never be experienced if we were deprived of our sense of touch. Touching people and objects is part of one's daily living. Though touch is an essential sensory modality for the attainment of cognitive and affective information, little more is recognized than its tacit role in socialization. It is necessary to study the processing of tactile information since it has been an important source of learning all through evolution. Although touch encompasses the whole body, the present manuscript will focus only on the hand and its fingers and the ways in which hands acquire information in touching and manipulating objects. The human hand with its multiplicity of functions is a versatile organ and more important, its association with the sense of touch with its fine sensitivity at the fingertips equips the hand for a variety of haptic processes. Sherwood Washburn (1960) insisted that the modern human brain came into being after the hominid hand became handier with tools, maintaining that the brain was the last organ to evolve. The versatility of the hand is evident in the ways in which it is instrumental in handling objects that are made to function as if they are an extension of the hand. The experiences of people that accomplish their pursuits of skill in handicrafts, sculpture or any other, are examples of the flexibility and variety of hand actions.

An attempt to review the development of tactual processing follows by comparing blind and sighted children in tasks that assess cognition by touch. The role of vision in cognition is questioned to reveal that vision is neither essential in the attainment of information and does not necessarily affect the general maturation of the child. The purpose of the manuscript is to generate awareness amongst readers about the importance of a relatively neglected sensory modality that is involved in most forms of development. Although touching with the hands provides convergent information such as feelings of pressure, temperature, size or shape of objects, perception and learning are also affected by other sensory modalities. Thus it is of relevance to examine some of the important processes involved in the acquisition of information.

1.2 Ways of knowing

The world about us abounds with information that conveys relationships between people, between people and objects, or between people, objects and events. The foremost experiences are sensory, as when an infant first opens its eyes to daylight or hears the mother's voice. Queries as to how the infant understands sights from sounds or relates the mother's face with her voice have posed an unending series of challenges in the quest for unraveling some of the fascinating puzzles of perception.

Attempts to study sensation and perception in infants have occupied the interests of developmental psychologists over the decades and have mainly centred about the theoretical discourses of Heinz Werner, Jean Piaget and Lev Vygotsky. These theorists were concerned about the ways in which human knowledge is acquired over periods of ontogeny and they followed a relational approach in studying theoretical constructs. The relational approach to the study of any phenomena treats dichotomies such as the individual/collective, nature/nurture, subject/ object, form/process, not as separate and distinct entities, but as lying on a continuum. The attributes at the two ends of the continuum are said to relate to each another in varying degrees. A split approach tends to dichotomize processes such as activity and inactivity as separate entities, and not lying on a continuum. The relational approach is inclusive and admits all aspects of transformation or variation that may come about between the two ends of the continuum and merges any existing dichotomy, so that the bipolar attributes relate to one another.

Questions as to how we acquire information about the world have initially been a part of philosophy. Metaphysics is the broad area of philosophical inquiry concerned with conceptual inquiry into the nature, origin and structure of the world or being. Ontology is the domain of metaphysics concerned with the question of what constitutes the real (Putnam 1987), and epistemology is about knowing, and its primary question concerns the validity of what we know. Both are related to one another, in that epistemology tells us about how we know what is real, whereas ontology explains the real as we know it.

The epistemological doctrines of rationalism and empiricism can be traced as far back as Plato and Aristotle. Rationalism claims reason as a foundation to knowing, and empiricism claims that the senses or observation are a foundation to knowing. For Plato and Aristotle there were no radical splits between ontology and epistemology. Each took the problem of knowing as his focus, with reason and observation, form and matter as being extensive for understanding the world. Plato favoured an epistemological emphasis on reason with the ontological significance of form and pattern being foremost, whereas for Aristotle, the dialectical balance of reason and observation was important and emphasized the significance of the relational nature of form and matter. Therefore ways of knowing that favoured a relational as against a split or an exclusive approach seemed to be promising. Some of the primary ways of knowing about the world or seeking information are from perception, action and language.

1.3 Perception

Perception is a mental process that enables one to understand information derived from the senses. Heinz Werner considered perception to be foremost in development. Werner (1957, 1958), one of the three major developmental psychologists of the last half of the twentieth century, drew his theoretical approach from the dialectical features of the relational approach. In this context, he proposed the orthogenic (normal development) principle as a universal explanatory principle, or law of transformational change. Transformational change signifies a process of evolution during which an individual progresses toward a change that alters the individual so that he is better equipped to transact in a changing environment. The alteration indicates a level of complexity that has been acquired by the individual. According to Werner, the orthogenic principle asserts that whenever there is development, it proceeds from an initial state that is relatively global with little differentiation to a state of increasing differentiation, articulation and hierarchic integration. Studies with infants have revealed that methodological advances in designing tasks that attract the infant's attention have contributed to the understanding of how infants get to know their environment. Behavioural measures such as an infant's preference for one of two stimuli referred to as the spontaneous visual preference (Fantz 1958, 1961), or the waning of attention to a repeated presentation of a stimulus, or the habituation response and subsequent attention to a repeated presentation of a stimulus (dishabituation), have been widely used in a variety of ingenious studies with infants. The successive use of these methods in infant studies has revealed the importance of attention as a primary process concerned with the selection of information from the environment. Infants as young as four and five months are able to discriminate shapes that have been placed in their hands (Streri 2005). This indicates that tactile perceptual abilities precede visual shape recognition and tactile information is assessed by the hands early in infancy.

Fantz (1958) stimulated increasing work with infants' preferences, and questions were pursued as to what aspects of the visual stimuli determine these preferences. Evidence for effects of complexity (Hershenon 1964; Brennan, Ames & Moore, 1966) or visual scanning (Salapatek & Kessen 1966) about the vertices of a triangle showing eye movements indicated that by three or four months, the infant is capable of fairly detailed scanning that affords the opportunity for effective encoding of stimuli and learning. The primary functions of vision seem to be the recognition of objects in the environment and their spatial layout (Marr 1982). The perception of depth is crucial to determining spatial layout. Classic studies with the visual cliff (Gibson & Walk 1960; Walk & Gibson 1961) have indicated some awareness of depth in infants. Yonas & Granrud (1985) pointed out that infants become initially sensitive to kinetic depth followed by binocular and monocular depth information. Perceptual constancies such as the shape and the size of the object are learned with the help of size and distance cues.

Auditory perception is the process that detects sounds in the environment and therefore provides another source of information. Studies with neonates indicate that the auditory system is functional at birth, but that the auditory threshold is higher in infants than in adults (Engel & Young 1969; Schulman-Galambos & Galambos 1979). The human voice contains the richest source of information and neonates have been found to be sensitive to it. Young infants within a week after birth displayed more head turning to the sound of a human voice (Alegria & Noirot 1987) and could be conditioned to suck to the sound of folk songs (Butterfield & Siperstein 1972) and to the voice of their mother (DeCasper & Fifer 1980). Furthermore infants at the age of one month are able to detect phoneme sounds such as p or b (Elimas, Siqueland, Jusczyk and Vigorito, 1971) and indicate the onset of categorical perception or distinct speech sounds, though language development is facilitated more by syllabic configurations that are heard in a context with other words (Elimas 1974, 1975).

During normal waking activity, information is available simultaneously from different sense modalities. For example, faces and voices are matched, just as the texture and colour of an object are matched and adults are able to predict the sound an object would emit by looking at it. The question as to how intersensory integration takes place has been explained by two major theoretical positions. The traditional view is that the senses are separate at birth and become integrated during development. Subsequently this view was challenged by theories that argued that the senses are coordinated at birth. Although initially there is a primitive unity of the senses at birth, the senses get differentiated during development (Bower 1974). Examples of the ability of infants to synchronize lip movements with nursery rhymes (Dodd 1979) as well as their reliability for looking at films that matched sounds (Spelke 1979, 1981) are indications of their ability to integrate information from two different modalities. Intersensory coordination in the latter part of the first year seemed evident when infants were able to reach for a toy they had felt and heard, and not seen before, while it was paired with another toy they had neither seen nor heard (Bryant, Jones, Claxton & Perkins 1972). Thus infants were able to correlate the attributes of the tactually felt object with the visually seen one. This is an indication of the ability of infants to perform the mental act of representing the tactually felt object and match it to the visually presented toy. The unity of the senses has long been presented by Gibson (1969). She emphasized that the senses pick up specific information to integrate meaningful perception of the object or situation. For example, there are several instances of an infant rejecting a feeding bottle when the milk is not presented in the usual bottle even if the new bottle visually resembles it, indicating that infants are sensitive to the feel of the bottle both within the hands and in the mouth. Therefore the ability to integrate sensory information is evident at an early age.

The ability to represent information about an object and recognize the object later signifies the onset of cognitive development. Piaget (1954) demonstrated the origin of representations in his fascinating studies of the object concept. According to Piaget, the young infant of 1–3 months of age attains knowledge about the environment by acting appropriately on the object of interest. This early period is termed as the sensory motor period during which infants respond to stimuli by action schemes such as reaching or turning toward the object. For the infant the object that is reachable or graspable is the object that exists and an object is known as a graspable object. The object is graspable so long as it can be seen in the vicinity of the infant. However, if the object is hidden or goes out of sight, the infant no longer searches for it and treats it as nonexistent. This is startling, and more startling is the behaviour of the infant who fails to search an object that has been occluded in full view of the infant. Even if at a later stage the infant is able to partially retrieve the object, it is only between the ages of nine to eighteen months that the infant successively begins to understand that an object that is occluded continues to exist to be reachable. This understanding on the part of the infant marks the end of the sensory motor stage and the proper onset of cognitive development. According to Piaget the infant has learned about the existence of an object by acting on the object with appropriate action schemes and this has enabled the infant to construct his world as one that contains permanent bodies or objects. Thus the onset of the object concept demonstrates the acquisition of knowledge. An infant that searches for his fallen toy has attained the knowledge that the toy exists below the cot and that the infant must use appropriate action schemes such as reaching toward it or even crying aloud to attract the attention of the care taker toward the fallen toy.

Sensory functioning and perceptual processes undergo continuous change with age. A gradual decrease in the efficiency of the sensory structures is accompanied with an improvement in the abilities to perform complex perceptual integration and discrimination. For example, though visual acuity decreases with age, visual illusions are also found to decrease with age implying an increase in perceptual learning. The information obtained from eye movements is used to reduce a perceptual error and this decrement is associated with age changes in the magnitude of the illusion. The fact that learning and experience can produce changes in individuals has led to the speculation that many changes with age are an outcome of perceptual learning.

Direct tests of the effects of learning on perception involve conditions in which the individual may be confined or restricted to particular situations, in order to know the effects of these conditions on development. For example in instances of restricted rearing, sensory motor learning, perceptual rearrangement or the effect of context on learning, the impact of particular experimental arrangements were assessed by the performance of the individual.

Von Sendon's (1960) studies on cases of congenital cataract that were later restored to sight are examples of restricted rearing in which people were unable to recognize familiar objects by sight. Patients that received vision after being operated for congenital cataract were unable to visually recognize objects they had previously felt without vision, indicating the importance of vision in the spatial representation of objects. Perceptual deficits were also observed in other instances of dark rearing with animals (Riesen & Aarons 1959). Newly born kittens were reared in dark rooms for a few weeks in the absence of any light. When the animals were later brought to daylight, they had difficulty learning to make visual discriminations of patterns, and displayed erratic or abnormal responses, to the depth or distance of targets. However, learning is not only restricted to the eyes, but could involve the entire body as in the instances of sensory motor learning. Some authors believe that normal perceptual development depends upon active bodily movement under visual guidance. Von Holst and Mittelstead (1950) distinguished between stimulus input that passively acts upon an observer, exafference and stimulation that changes as a result of an individual's own movements, reafference. Animals deprived of normal movement in the environment (Held & Hein 1963) displayed deficits in activities involving depth perception such as dodging a rapidly approaching object, or avoiding the deep side of the visual cliff and failing the visual placing response which is a paw extension to avoid collision. Tests of perceptual rearrangements have indicated that our perceptions of space and direction are learned effects and such learning is facilitated by the notion of gravitational direction of events in the environment such as smoke rising or water falling (Kohler 1964). Appropriate visual guidance is an important factor for adaptation. When perceptual rearrangements were induced by requiring subjects to wear prisms, the location of objects viewed through the prism goggles shifted to the direction of the apex of the prisms. After several testing trials, the subject was able to assess and correct the error committed during the perceptual rearrangements, and adapt to the distortion of the prism. Held and Hein (1958) argued that adaptation depends on active movement just as the development of visually guided behaviour.

Context and meaning also provide a background for learning and affects our perceptions. In psychological studies, observers recognized unfamiliar or broken figures that cannot be easily recognized on paper, when these figures were presented in the context of a story (Steinfield 1967). Language and expectation can also modify percepts (Buckhout 1976). Attaching verbal labels to outline figures can influence subjects to categorize response items. The context provides cues that facilitate learning by allocating items with verbal labels to category memberships (Bousfield et al. 1958). In their classic study Bousfield et al., showed that subjects recalled names of vegetables and fruits in an organized manner from a randomly distributed list, because the context of recall (vegetable/fruit) gave meaning and structure. Thus, perceptual learning provides an unending source of information that gets modified with experience during development.

An example of perceptual learning is the choice of using the fingers for eating. In eastern cultures of the past where rivers were many and water was plentiful, the hands were washed before and after meals. The food was finely cut, whether vegetable or meat, so that it can be held in the fingers before it is raised to the mouth. Eating with fingers is considered to be a skill. For example Indian foods such as chappati (flat bread) or rumali roti (handkerchief like bread) can only be eaten with the fingers as the bread is dipped into one of several varieties of finely cut and spiced meat or vegetables. Western cultures of the past have lived through ice ages when water was scarce. This created the need for using cutlery during meals rather than raising freezing fingers to the mouth. Also the need for carving knives and forks are a necessity if large pieces of roast meat are to be sliced, particularly when vegetables are sparse. Therefore necessity created the invention of cutlery, and the choice of eating with fingers or cutlery is a matter of perception and learning. The choice of eating with fingers rather than using cutlery is a good example of the effect of context on the perception and learning of appropriate actions that have been performed and sustained over centuries.

1.4 Action

Jean Piaget explained his classical theory of equilibrium to be a consequence of transformational change in the developing child. Piaget followed a relational approach in which he attributed the child's process of constructing his environment as requiring him to maintain equilibrium between assimilation and accommodation. Assimilation is the process of attaining information, such as observing and knowing that the water level in a tall and slim jar is higher than the water level in a short and broad jar. Accommodation is the process of adaptation. The child has assimilated the water levels in the jars and is able to adapt to the situation by accommodating the differences in the water levels of the jars as a consequence of the different shapes of the jars and understanding that in actuality both jars contain the same amount of water. Both assimilation and accommodation seem to proceed from a state of non-differentiation to a state of differentiation. According to Piaget, action or action schemes are fundamental for the infant in the acquisition of information. The child according to Piaget constructs his environment by acting upon it with structures that get modified during development. For example, a child between the ages of 4 and 7 years may have the ability to represent objects, though his ability to think about events or situations may be confined to a single feature or characteristic about the event. If presented with a tall jar and a flat jar that contain the same amount of water, the child will conclude that the tall jar contains more water. Thus the child considers only the height of the water in the two jars and does not consider the width of the two jars. Such a child according to Piaget is at the preoperational level and does not possess the ability to operate simultaneously about the different characteristics of the situation. An older child of 8 to 12 years at the operational level is able to simultaneously assimilate and accommodate the different characteristics of the situation and understand that both jars contain the same amount of water. However, this child relies on the concrete characteristics of the situation and may not be able to solve a hypothetical problem in a similar situation. It is only at the formal operations level or after the age of about 13 years that a child would be able to solve hypothetical situations in the absence of any physical or concrete props. Therefore development proceeds through the process of unraveling or decentering from the central features of an event such as the height of the water level in a jar to including the width and other features of the jar, from irreversible rigid behaviours such as the inability to understand that a round ball of clay when elongated has the same content, to reversible flexible actions when the child understands that the same ball of clay can take different shapes; from understanding static states to transformational states of the same content be it clay or water, all being an outcome of actions that have culminated in mental constructions that indicate qualitative changes in behaviour.

The development of a child as s/he progresses in the active construction of his environment is an indication of the contrasts between early childhood and middle childhood cognition. The younger child's approach to a problem situation is more global or qualitative, whereas the older child seems to understand that problems can have precise, specific, potentially quantifiable solutions. Thus according to Piaget, a child learns by acting upon the environmental situations and this enables at a later age to think and solve problems hypothetically in the absence of any concrete or visual props. There are several examples of family occupations in India such as pottery or carpentry that can affect the conservation of quantity in young children who belong to families of these occupational groups. For example, it has been observed that sons of potters attain the conservation of mass at an earlier age as compared to their school going peers because the manipulation of clay during pottery making, enriches their concepts of the quantity of mass (Kumari 1976). Similarly, sons of carpenters have been observed to attain concepts of measurement at an earlier age as compared to their school going peers because of their increased familiarity with measuring instruments of carpentry in the family. Actions of children in appropriate occupational contexts facilitate conceptual development and knowledge of particular occupational skills.

As the child grows from infancy to adolescence, her/his social knowledge develops in a predictable age related manner, that is, s/he makes progressive reorganizations in the principles that constitute the various aspects of social knowledge. Children first notice observable characteristics- the appearance and behaviour of themselves and other people. Soon after they become aware of internal processes- the existence of desires, beliefs, intentions, abilities, and attitudes. Social understanding as well gets organized with age, as children gather together separate behaviours into an appreciation of their own and others' person-person behaviours from simple one- sided explanations to complex interacting relationships that take into account both person and situational variables. Finally, social cognition moves toward a metacognitive level of understanding. That is, as children get older, their thinking is no longer limited to the social reality of the situation, but they also think about their own and other people's thoughts. The development of social cognition is age related (Harris 1989). Each stage in development is accompanied by new relationships that are formed which differ qualitatively from previous relationships. For example, an infant's attachment to a parent is expanded to include the extended family, peer relationships and society at large. In addition, a successful relationship in one context will contribute to successful relationships in other contexts, indicating continuity in social development. As they get older, children become aware of multiple sources of information for inferring emotion. Children's experiences of emotion lead toward profound and pervasive changes in their social cognition and social behaviour.

There have been many investigations of children's understanding of mental representations. The majority of these have dealt with children's comprehension of representations that differ from one another or from reality. Experimenters usually test children's understanding of beliefs by using false belief tasks and the common finding has been that three year olds tend to fail these tasks. In contrast 4 and 5 year olds usually pass such tests showing that they have an understanding of another person's mental state. However, it is not until middle childhood and later that children appear to gain any appreciable understanding of the mind as an interpretive, constructive processor. Many investigators believe that the core knowledge that is developing in this area can be characterized roughly as follows: People mentally represent/ perceive/ interpret the world as being in a certain way. However, their mental representations are not always accurate and can vary from person to person or vary within the same person over time. For example, their present belief about something may be false, or it may differ from someone else's belief or from their own at a previous time. How things appear perceptually may differ from how they really are or from how they appear from a different viewing position or perspective. The child's understanding of knowledge is complicated. By the age of 4 or 5 children seem to have some sense that knowing is a mental state that is experienced as certain and refers to a true state of affairs. Correlational evidence suggests that an understanding of knowledge, certainty, false belief and appearance reality all develop together, perhaps as different manifestations of a developing theory of mind (Moore & Furrow 1991). Studies that have tested young children's understanding of the importance of perceptual access in acquiring knowledge found that even three year olds will tend to attribute knowledge of a box's contents to a person who looks inside the box than to one who touches the box (Pillow 1989; Pratt & Bryant 1990).

With the development of social cognition, an aspect of the child's capacity that needs to be understood is his/her ability to construct society; where the child places herself within the societal framework and how she develops an understanding of her role in it. Between the ages of 18 and 30 months a number of psychological acquisitions emerge in the child. These include the speech of others in her society and her subsequent ability to enter into a narrative mode, the mutuality of peer relations such as the existence and recognition of another child, fantasy and imagination as in wanting to be 'Superman', solitary and pretend play like pretending to drink hot tea from an empty cup, complex emotions such as bad and sad and a sense of morals as to what it is to be a good child. A psychological prerequisite for these accomplishments during development is the competence to represent mental objects and the desire to form mental symbols. According to Piaget (1954), the cognitive construction of these mental objects begins at about eighteen months. Children become capable of constructing symbolic mental material in a variety of activities. The question arises as to whether there is a unifying framework that encompasses these activities apart from symbolic competence. According to Furth (1994), there exists such a frame since all activities imply an inherent grasp of a societal frame on the part of children. For example children recognize the role of their parents and grandparents. This grasp goes well beyond the social familial interpersonal context in which children develop since birth, and beyond what they could learn from outside models such as peer groups or teachers. Furth claims that children for the first time can have a sense of the collectivity of a human society when they become active participants and co-constructors of a societal frame. The child understands that s/he could be a daughter at home, a friend at school, a student in class and a member of a school/group.

According to Furth (1997) there is an endogenous origin of a societal frame, or an instinct for society. The suggestion here is that the societal institutions come to the child not merely from outside as an exogenous learning as proposed by social construction, but begins from birth in so far as everything connected with infant care is influenced and directed by cultural forces. To this extent, the child is a passive participant of society. But with the developmental emergence of mental objects and the signifying symbols, the child's position vis-a-vis society changes dramatically. The child develops the ability to represent people, objects and events in their absence. At this point in development children become active as endogenous constructors of a societal frame. The child is able to imagine future actions. The point at which the psychology of the child and the society meet in this perspective is in explanation of mental objects. These objects are mental constructions of the child and are experienced as imaginary events, stories, fantasies or images. The claim here is that these mental constructions, all referring to objects, have an inherently societal character such as the need to play and win, or lead a chorus of singers, or develop a friendship, or build a career, such that the mental object itself is a societal object.

Although children's aspirations have been little studied, there are endless examples of a child wanting to grow up to be an artist or a sailor. Children also pretend sailing in big seas or imagine being a Hussian or Picasso. These are mental constructions of children that display their ideal societal character. Like Don Quixote, every normal child wants to follow her/his dream_ a dream that will make her/him acceptable and recognized in society. To this end the child's fantasy about the future enables her/his creativity and productivity in the area or subject of choice. The actions are achievement oriented and the created product is a reflection of her/his mind. In Furth's studies, the children were amongst themselves and not under the direct constraint of the adult. The children's actions were therefore likely to be motivated by self- regulation drives and motives from within such as wanting to play or wanting to draw with crayons, rather than in response to an externally set task. This, according to Furth is what is meant by the claim of an endogenous origin. The notion of endogenous origin is in conformity with Piaget's use of the term for the development of human logic. Piaget presents it as a third choice that involves action, desired and spontaneous, while rejecting both innateness and learning from outside. In this theory, development in contrast to physiological maturation and exogenous learning means endogenous construction or co-construction of the child's thoughts and actions, like the choice of preferred colours in drawings or building blocks at play.

The emphasis in Furth's explanation is that knowledge of society is not innate, but is a developmental construct. In this constructive sense, it is claimed that long before each different culture has succeeded in socializing its children, they already live in a societal frame such as the child and the caretaker. As soon as children are able to construct a mental object and articulate a mental desire, the object and the desire already include society. The child gets the bicycle s/he desires. The creativity afforded by the logic of the object is spontaneously directed toward the instituting of society (my new cycle) just as the desire of the object makes the child spontaneously attached to society (my gift cycle). With the development of a symbolic mental life around two to three years of age, face- to- face inter-individual relationships and interactions spontaneously take on a new, societal quality. Phenomenologically, children begin to live in a societal world. On account of the developmental transformation beginning around two years of age, the child's action communication becomes societal language, the child's individuality becomes the societal role of a girl or a boy and the role of a child belonging to a particular adult society. The child's learning and doing take on the role of a particular culture and the child's pleasures and emotions become attached to specific values and ideologies. The child begins to have an understanding of the self as an agent. The child is able to express his liking: 'I like ice cream more than toffee'.

As children develop an appreciation of their inner mental world, they think more intently about themselves. During early childhood they begin to construct a self concept which consists of one's beliefs about oneself- one's appropriate roles, personal attributes, worth and values. The self concept may emerge in a subset of occasions of shared attention_ namely those in which the object of joint interest is the child herself (Tomasello 1993). These are on occasions when the mother says 'that's a good girl', or 'did you do that'? when she speaks to the child about the child. It follows that the child like her mother now takes herself as an object of thought. She begins to think of herself as having traits, attributes, worth and value. The age of onset of self recognition is culturally dependent and varies with parenting styles (Keller, Yovsi, Borke, Kartner, Jensen & Papaligoura 2004). Though infants as young as three months have displayed self recognition with distal (e.g. eye contact) and proximal (body contact) parenting styles, mirror self recognition is not always understood in very young infants (Keller, Katner, Borke, Yovis & Kleis 2005).

The child's actions in a multiplicity of contexts provide occasions for learning and knowing about the environment. Most often it is the integration of sensory information that sensitizes the child to a particular action or situation. For example among children in artisan families, the traditional art and craft, whether carpentry, pottery, sculpture or jewellrey making, is taught to the child at an early age. The need to be proficient in the family craft is instilled in the child as soon as the child is considered to be mature to perform and receive instruction. Learning the traditional skill or craft involves action and movement of the hands and fingers that are guided by motor planning. Motor planning, however, is not a unitary quality, but a collection of abilities. Motor planning improves with age as children improve in their ability to observe and replicate movement series of various types and complexities. Proficiency in motor planning may also be due to number of variables: (1) the ability to attend to the prolonged demonstration of a movement, (2) the ability to organize and remember what has been demonstrated, (3) the maturation of the integrative processes, and/or (4) the ability to execute a series of movements correctly (Cratty 1982). Children's observations of the actions of the elders, stimulates their need for skill attainment. Therefore, actions motivate learning and achievement.

1.5 Language

Vygotsky (1962) maintains that development is best characterised as a complex dialectical process that is accompanied by varying periods during which the child undergoes transformation. For Vygotsky, language constitutes the basis of socialization and development. The infant, according to Vygotsky, is exposed to information from others and manifests his mental state through language acquired from others. It is only later in childhood that the child becomes egocentric and reflects on the use of language for socialization and communication. Piaget (1953) claimed that the child is initially autistic and he drifts into egocentric speech to externalize his thoughts. Uncomfortable with Piaget's conclusion that early speech is generated asocially, Vygotsky in a series of experiments demonstrated that a child talked twice as much when presented with an obstacle. This externalized conversation with oneself, commenting on and predicting the results of an action, is the natural dynamic of problem solving, and this talk is sensitive to social factors. Piaget had likewise observed that egocentric speech occurs more often in a social context and that the child assumes that he is being understood. However, Vygotsky observed that when the child is placed with others in a room his egocentric speech drops to one fifth its previous rate. Vygotsky concluded that egocentric speech was not, as Piaget had suggested, a compromise between primary autism and reluctant socialization, but rather an outgrowth of speech that had been socially and environmentally oriented. The developing child was not externalizing his internal thoughts, but internalizing his external verbal reactions. Vygotsky concluded that development in thinking is from the social to the individual and not from the individual to the social.

As a reaction to some of the traditional approaches to socialization (Bullowa 1979; Schaffer 1977), the nature of preverbal communication has been considered to be an important condition for the origin of language. It has been argued that language as a system of communication is grounded upon modes of relating, already functioning prior to the onset of speech (Lock 1978). Most of these studies assume the primacy or dominance of the visual system as the means through which the child comes to know the world. Both mother- infant interaction studies and studies concerned with the origins of language have concentrated on play exchanges observable between mothers and infants in the first three months of age, in which eye contact is given a central role in the regulation of the interaction (Bruner 1977; Trevarthen et al. 1975). They comment on the role of gaze direction in the monitoring of mutual attention to objects, a capacity which is said to provide a basis for establishing a shared frame of reference (Bruner 1977) and the use of gestures such as pointing and reaching, which along with offering objects and give and take games, initially emerge in the last quarter of the first year (Sugarman-Bell 1978; Trevarthen & Hubley 1978). Initially accompanied by prosodic marking, gesturing persists once the child begins to use words to convey particular communicative effects (Brunner 1978; Carter 1978). This persistence is taken as evidence for continuity between preverbal and verbal communication.

Gesture is used in teaching contexts (Flevares & Perry (2001); Goldin-Meadow, Kim & Singer 1999) and promotes learning (Church, Ayman-Nolley & Mahootain 2004) because children are likely to profit from instruction that includes gesture by understanding the words that accompany gesture. Listeners are able to better understand the message conveyed by a speaker when the message is accompanied by gesture (Goldin-Meadow & Singer 2003). Children produce substantive gestures when they are required to reason about objects than merely describe the objects (Alibali, Kita & Young 2000). When children were exposed to gestures of their teachers as they gave instruction to children in both speech and gesture, the children reproduced the information conveyed in the instruction more often in their own gestures than when given instruction by speech alone (Cook & Goldin-Meadow 2006). Cook and Goldin- Meadow concluded that gestures are important for learning because they convey information visuospatially and are therefore able to clarify aspects of a problem better than speech alone. Gestures also use the body to represent actions by using the hands to display signs of addition or subtraction along with spoken numbers, and these embodied representations are involved in cognitive processes such as working memory (Nilsson et al. 2000) and linguistic processing (Zwaan, Madden, Yaxley & Aveyard 2004). Furthermore gestures are assumed to reduce the load on working memory because speakers remember more items both verbal and visuospatial when they gesture than when they do not (Goldin- Meadow, Nusbaum, Kelly & Wagner 2001; Wagner, Nusbaum & Goldin-Meadow 2004).

The importance of social factors in the development of thought and speech is not only evident in normal sighted children that respond to gestures and facial expressions of mothers or care takers, but also among totally congenitally blind children (Urwin 1982).

Urwin (1982) emphasized the role of the social context in the development of language. She demonstrated the effects of such socialization in congenitally blind children that were delayed in language acquisition (Wood 1970; Wills 1979). Urwin indicated that the choice and use of single words by children during play or in the presence of others is an out -come of their social interaction. She argued for the social origins of representation in language and questioned the assumption that social and cognitive processes make separate contributions to the emergence of language.

Among hearing- impaired children social factors such as parenting, type of school, oppourtunities for play and communication by signs have been found to affect their neurodevelopment (Sharma 1998). In fact the reason for the major differences found between the hearing and the hearing- impaired children in perceptual motor and laterality tasks have been attributed to the differing social conditions in the two groups. The hearing impaired children as compared to their normal cohorts had delayed balance and motor development. Since the early 1900s research has consistently identified differences between deaf and hearing children in the performance of a wide variety of motor tasks, most notably balance (Boyd 1967; Brunt & Broadhead 1982; Butterfield & Ersing 1986). Balance task performances are dependent on three major subsystems (a) vision and its ability to stabilize the visual perceptual field of the child, (b) movement qualities involving the precise correction of imbalances, and (c) the mechanisms of the inner ear that act as integrators of visual and motor information impinging upon the maturing child. Defects in any of these systems may produce balance problems for the maturing child that in turn will impede the stability needed to exhibit new and emerging skills of the larger muscle groups including kicking, throwing and catching. This may also affect their perceptual motor abilities (Sharma 1998; Ittyerah & Sharma 1997). The planning and execution of motor skills involves the hands and limbs. The development of motor planning abilities does not proceed in the same manner as in all children. Motor planning problems have been related to motor planning behaviours, speech language functions, and cognitive intellectual qualities (Kimura 1982). Motor planning skills have been known to involve both verbal and task type components (Cratty 1982). The hearing impaired children have tendencies for delayed motor development in tasks that involve language and motor components (Ittyerah & Sharma 1997) indicating the role of language in thought and development.

The increased tendency of the hearing- impaired to use gestures that have been observed visually, affects preferences for exploring structures that are similar to basic sign language. This is true for deaf children of deaf parents where sign language is the linguistic vehicle. These children acquire sign language in essentially similar ways and through similar mental strategies as hearing children acquire spoken language (Bellugi 1988, 1991; Petitto 1983, 1991). Sharma (1998) observed that her hearing impaired children showed a different pattern of hand use in signing words. This may be explained by the fact that gestures and language are two sides of a single system. Gestures and language form an integrated system in which semiotic properties are connected. There are a number of linkages that imply that gesture and speech should be viewed within a unified conceptual framework as aspects of a single underlying system. Gestures and speech develop together in children; gestures and speech break together in aphasia. These could be some of the reasons for the delayed language and different hand use in signing words in the deaf children studied by Sharma.

Children's first gestures are concrete pointing at objects or people. As children's language moves from a largely referential focus, through descriptive elaboration to the ability to structure discourse, so their gestures develop from a largely concrete deictic emphasis to the discourse referring gestures. Despite the fundamental character of the differences between gestures and speech, they are closely tied together in meaning, time, function, development and dissolution (McNeil 1992). Thus Sharma's findings lead one to believe that language emerges primarily from mental representational capacities and children may well be attaining many language skills that are relatively domain or situation specific (Fischer & Corrigan 1981). Sharma (1998) observed that in tactile matching tasks with the hand and fingers, the normal children exhibited a left hand advantage for all tasks, whereas the hearing impaired had a left hand advantage for the recognition of form only. For all the other tasks (size, balance and texture), the hearing impaired had a right hand advantage. This may be explained by the fact that the finger quality improves from six years onwards in children. Tasks involving finger opposition and those requiring the child to touch a single finger repetitively to the thumb, improves at six and seven years. Overall, superiority in the preferred hand is seen more often in tasks that require successive touches of the same finger while sequential finger movements are often performed equally well by both the left and the right hands. Both hemispheres seem to contribute to repetitive and apparently purposeless action, whereas when the actions involve sequencing and an obvious objective, it is possible that there is a greater contribution by the left hemisphere (Cratty 1982). The differences in the laterality patterns may suggest that deaf individuals develop different neuroanatomical organizations than do the hearing (Neville 1988). Sharma's study may hint an association between cerebral lateralization and perceptual motor abilities and language tasks. Although there were no differences in the handedness of the two groups, balance tasks have indicated differences in the laterality patterns of the two groups. The development of perceptual motor abilities may be related to the maturation of the sensory motor cortex. As language functions are coordinated by the left hemisphere, the gradual maturation and development of the sensory motor cortex may result in the better acquisition of fine motor tasks.

Sharma's study also highlights parenting styles, communication with parents and communication with peers in influencing the motor repertoire of deaf children. If parents are not visually stimulating, they may not be interesting to deaf children, therefore showing less responsiveness in the children. Effective communication may indicate that parents may be highly stimulating, encouraging and instructing in developing motor skills in their deaf children, in order to maintain dyadic relationships between the parent and the child. The interaction problems of the deaf may result from their lack of knowledge of appropriate social and communicative strategies from inexperience, and the difficulty that two equally unskilled partners have in coordinating their communication with each other's attention.

Delayed language development, relatively poorer performance in perceptual- motor skills and different patterns of laterality as indicated by the hearing-impaired children, may be explained in terms of a contextual interactionist model that focuses on multilevel bases of organization for behaviours that are interconnected. Further, factors such as type of schooling, physical education curriculum, parenting styles and oppourtunities for play may lead to differences in motor development (Butterfield et al. 1993). All these may also contribute to varied motor experiences resulting in differences in motor planning skills and in turn influence the performance of the deaf on a number of tasks as well. Thus social and cognitive variables combine during development to promote the required context for the acquisition of knowledge.

1.6 Reading with fingers

Just as the deaf acquire language capacities by signing and gesturing with their fingers, the congenitally blind individuals learn to read with their fingers. In this manner, perception and actions of the hand and its fingers serve to communicate through language both spoken and written. Reading by touch involves the process of pattern perception of tactual symbols. Shape perception by touch is an intersensory process because the spatial reference cues depend on the balance of inputs from touch, movement and posture and that varies with the size and composition of objects. Braille script is derived from a composition of six raised dots that can be felt on paper with the fingertips of the reading hand. The perception of Braille patterns is regarded as a constructive process, which starts from the detection of dot density disparities by relatively unsystematic scanning movements (Millar 1997). The scanning movements of the fingers indicate the versatile nature of the hand.

Braille reading depends on the ability of the fingers to be able to perceive the shape of the characters. Tactual shape perception is often tacitly regarded as inferior to vision, because shape perception by touch takes longer and is less efficient than vision. Discrepancies between findings for vision and touch for reading are attributed to differences in sensory acuity and to the fact that vision can take in more information in one glance than is possible by one touch (Millar 1994). However the kind of information that is perceived by vision and touch is not the same. Though the visual and tactile sensory modalities may be comparable for shape perception of objects, other aspects of the stimulus such as its texture and weight, can only be accurately perceived by touching or holding the object.

The composition of Braille is an important factor in understanding the processes that underlie its perception. Braille characters are all derived from a six (2×3) dot matrix, called the Braille cell. The cell is approximately 6.3 millimeters high. The characters are small. The small size of the Braille patterns is frequently considered the main limit on processing, because it presents problems for tactual acuity (Millar 1997). As in vision, detecting the presence of a pattern is determined by the contrast between figure and background. In touch this depends on the height of the raised dots of which the figure consists. Louis Braille who was blind himself and was educated as a blind person invented the Braille system of reading in the eighteenth century. Since, Braille has proved very successful in that it is still the main system of written communication for blind people all over the world.

Braille involves active scanning movements by the fingers of the hands. During reading, the speed and movement of the fingers and the pressure that is exerted, are determined by the reader. Movement cues are important for the recognition of Braille patterns and enter the acquisition of reading from the start (Millar 1997). There has been considerable speculation about the functions of the two hands, when both hands are used and how they may relate to fluent reading. Millar (1987) observed that the hands read in intermittent alternation. When the right hand read the last section of a line, the left hand was engaged in finding the start of the next line. When the left hand was reading new text, the right hand moved to join it. The division of labour between the hands during reading Braille was between the verbal and spatial domains. That is, as one hand feels the text for words the other hand feels for spaces between words or letters. In fluent reading, the advantage of two- handed reading thus seems to be in the division of labour between the hands for verbal and spatial processing domains. In two handed reading, the right hand mainly reads the right hand portions of the line and left hand reads the left hand portions of the lines. When the mid point of one forefinger is touching a letter, the other is most often either on a blank between two letters or between two words (Millar 1987). Furthermore, the hand that people use for reading Braille has little or nothing to do with their general laterality (Ittyerah 1993). Although one may have a hand preference for eating or writing, each hand is as able as the other in tasks that require tactile ability. Thus the hands are versatile organs that combine inputs from touch and movement to enable perception and understanding of information.

1.7 Sensory modes of concept attainment

The acquisition of concepts is an outcome of the interaction of the child with its caretaker and the environment. The child understands that animals and objects are real whereas stories may not be real. Gradually the child develops an understanding of what is real and what is not real or fantasy. Attempts to study fantasy/reality distinctions in children (Harris et al. 1991) have indicated that young children are able to distinguish real objects from imagined objects and this ability is an indication of the developing theory of mind. However, this may not be the case for congenitally blind children. We (Ittyerah & Goyal 1997) have shown that viewing cognitive development in terms of theory formation may not explain the development of blind children completely.

Synesthetic abilities may assist the congenitally blind to distinguish between contents of fantasy and reality. It was conventionally believed that young children are unable to distinguish systematically between mental and real phenomena. Subsequent research has led to a re-evaluation of this claim. It has also led to a debate on how mental states (fictional) like dreams, etc, compare with epistemic (dealing more with facts) mental states such as beliefs. To explain the origins of mentalistic stance in each person, developmental psychology has tried to find how and when young children acquire common sense mentalism. A number of theories attempt to explain children's understanding of the mind. For convenience they may be divided into two broad categories, though both types of accounts subscribe to the view that children can appreciate the experience of others as a consequence of cognitive development. The first set of theories (Astington, Harris & Olson 1988) deal with various phenomena such as false beliefs, etc, without going into what might have led the child to such phenomena. The emphasis of these theories is that the child is able to conceptualize others' mental states and thereby adopt a theory of mind that enables one to understand others. The second set of theories deal with the view that mental phenomena are based on publicly observable events that support interpersonal behavior (Hobson 1991). Children perceive persons and objects from interactions of touch and eye hand co-ordination. These have an affective component in that they may be reinforced. The interaction between the child and another enables the child to conceptualize the mental and physical existence of the other. Knowledge of these shared interactions is from publicly observable behaviour.

A question of related interest is to know how children build theories of mind in the absence of any visual information. Normal children learn from sensory interactions such as associating sights with sounds or smells. Children are able to recognize people and objects by sight and understand fantasy/reality differences. Hence vision has an important role all through development. Are the congenitally blind children able to make fantasy reality distinctions like their sighted peers? If they can, then would their concepts of less concrete events such as dreams or fairies have the same reality status as actual events such as darkness or space? The weight of the evidence suggests that there are some differences between the blind and the sighted in the organization of mental functioning in the early years, but for the most part the blind are able to gain information from other sources than sighted children. If so would their representation of make believe or fantasy differ from that of the sighted children? Testing these dimensions may provide potential information about the ways in which representation of knowledge and need to visualize are integrated.

Ittyerah and Goyal (1997) tested congenitally blind and sighted children for fantasy/ reality distinctions of real and imagined objects and the development of concepts of darkness, hiding objects, space, dreams, emotions, facial expressions, size, and height.

The study compared children's ability to hold mental images in the head with propositional imaging as in the stories that were used to test the eight concepts. Both groups of children were able to understand queries that relate to considering objects in the head as fictional and not real and also differentiate this exercise from images that arise in the story concepts outside their heads. The conceptual understanding of the blind and sighted children was similar for most of the stories, except for that of dreams and hiding objects for which the sighted had more reality responses. When the blind children were questioned about their dreams, they replied that it was about what they felt or heard and had no knowledge of colour. Since dreams have a visual quality about them, it is possible that congenitally blind children may not be able to integrate the information available in dreams to that available in a wakeful state and therefore may be unsure of the reality status of dreams. Even sighted children have not always been capable of offering appropriate answers to questions bearing on the concept of dreams. Laurendeau and Pinard (1962) reported that children considered dreams to be made up of material content such as wood or cloth. This indicates that they have difficulty associating a dream as a mental experience and consider such experiences to be made of materials that can be seen or felt.

For the concept of hiding objects, blind children were unsure of what it was to place an object before another and occlude it, so that another person is unable to see the occluded object. Therefore they may have a poor understanding of what it is to hide objects in reality and this is evident in the responses to the story concepts on hiding objects. The blind children were unable to understand that hiding an object prevented it from being seen by another. This is because lack of sight impedes perceptual information predominantly inferred by vision. This indicates that blind children's experiences arise from interaction of the self with others and is not implicit knowledge as the theory view suggests. Research indicates (Wolley & Wellman 1992; Estes, Wellman & Wolley 1989) that children by the age of three or four have a basic understanding of the unreal, non- perceptually public, internal nature of fictional mental states such as dreams and imagination and resolve the various dualisms such as real- unreal, public-private, internal-external, involved in coming to a mature understanding of dreams (Kohlberg 1969; Laurendeau & Pinard 1962). In general it appears (Ittyerah & Goyal 1997) that young blind children can successfully distinguish between fantasy and reality. The findings also suggest that the theory formation view, according to which cognitive development may be explained in terms of theory formation, may not explain the development of blind children completely. Since the congenitally blind have a sensory deficit, their dependence on alternate sources of information is increased, and hence their development is not entirely similar to that of the sighted.

Proponents of the theory-theory view (Astington, Harris & Olson 1988) believe that children have theories about others that are conceptually modified. However, according to Hobson (1991) children come to know that people have minds and such knowledge is acquired through the experience of personal relations that are biologically inculcated such as mother-infant interctions. This information provides perceptual-affective awareness of interpersonal and intersubjective co-ordination. Thus the nature of interpersonal relationship affects the acquisition of information. For example, Mead's (1932) observations about the absence of animistic thought among the younger Manus children indicate that animistic thinking is an outcome of adult beliefs that are publicly shared with their children in later childhood. Similarly if blind children have not had an understanding of what dreams are or what it is to occlude objects, it is because adults have not sufficiently shared or explained their experiences of these states with their children and not because these children are incapacitated in theory formation. For example most children are familiar with the stories of the Man in the Moon though all of them may not have seen him. Most sighted children may have seen 'the man in the moon', but blind children know of it from publicly shared observable knowledge by interaction with others. The blind may have an understanding of concepts or others in view of their own subjective orientation towards the world and they are not a part of implicit theory formation. According to Furth (1997), young children over time co-construct cultural know how on the basis of their interaction with and observation of other members of society.

In summary, touch yields information different from vision. In the absence of vision, blind subjects perform most tasks as well as their sighted controls, by relying mostly on self referent information. Therefore the question of what is relevant to vision and its role in the development of cognition remains to be known. In an ever growing society as India, with a fairly high proportion of educable blind people, attention needs to be directed toward a vocational future. If the blind are to benefit from training, their education must incorporate a vocational component that can be pursued from an early age, such as the utilization of both hands and practice at tasks that require tactile sensitivity. Though vision may provide external references for the sighted, the blind are found to use self- reference cues during performance and their visuo-spatial proficiency improves as well (Liben 1988; Millar 1994).

1.8 Queries and concern

Queries about the perceptual and cognitive development of the visually impaired children were concerned with comparisons of systematic accounts of early cognitive development in normal sighted children. Comparisons of interest were particularly with Piaget's sensorimotor theory and Werner's comparative psychology of mental development which emphasized the importance of eye- hand- mouth coordination and the active manipulation of objects under visual control for acquiring concepts of object permanence, (that is, an object continues to exist even if it cannot be seen), objective space such as distance and direction between two people or objects and physical causality that explains the causes of events as to why it is hot when the sun shines brightly. The understanding and application of these concepts emerge from effective contact with the environment. In this manuscript an attempt is made to explain that object permanence and concepts of space can be understood by blind children in the absence of vision. Some of the queries raised by Peter Wolff (1966) about a blind child's acquisition of the object concept and modes of sensory experience used to understand the environment are of immense interest in knowing about the attainment of knowledge in the absence of vision. If blind infants use different perceptual modes to learn concepts of the real world, it would be of interest to compare the differences in the styles for cognizing the environment in the blind and sighted groups. Studies of cognitive development in infants with congenital limb defects due to thalidomide poisoning indicate that normal concepts of cognitive development such as conservation and classification of objects are understood in the absence of physical contact of the object with the fingers or the palms of the hand. Further comparisons of a blind and sighted twin in the tactile discrimination of objects indicated (Wolff 1966) that the blind twin displayed good progress as compared to its sighted twin in the way he approached tactual exploration. Studies have shown that children who have never had sight tend to approach objects with a whole hand grasp. Sighted people exploring an object hidden from view tend to outline the object with their fingers. The blind twin however showed unusually advanced discrimination in the way he explored the length and shape of a wooden spoon. This does not indicate that blind infants or children have better tactual acuity than the sighted, but rather confirms subsequent findings that early in infancy the tactile sense is used more often than vision till the ages of three or four months after which vision begins to dominate (Streri 2005). By the age of six months eye hand coordination dominates most interaction and contact with objects and the reliance on touch decreases among sighted infants. The manuscript attempts to clarify that touch is the earliest modality to develop, but the least understood because its importance is ignored in most sighted conditions. Nonetheless the importance of touch in the attainment of knowledge is imperative and its effects in perceptual learning and development are pervasive for most transactions with the environment.

There are suggestions (Wilson 1999) that appropriate hand use is an indication of brain development and items that assess hand actions should be incorporated into intelligence tests that assess children and adults. This will also increase an understanding of the repertoire of nonverbal behaviours that can be used as items to assess cognitive development. There are several instances of nonverbal accomplishments with the hands such as playing musical instruments, sculpturing, carpentry or making pottery to name a few. An example of the versatility of the hand and fingers is the form of finger signs or 'mudras' as a form of language in Indian classical dance. Mudras formed by the fingers of the dancer communicate different mental states, such as joy, sorrow, fear or anger and are synchronized with choreography to display the art of dance forms before an audience. The state of emotion displayed by each mudra is an indication of the strength of communication that the dancer intends to share with the shapes and forms of her/his fingers. These dance forms have been sustained over centuries in their variety and richness by the many finger mudras of many a dancer that has effectively communicated tales of history and mythology to audiences far and wide.

In this chapter an attempt has been made to organize the ways in which people learn about the world around them. In principle the main ways of knowing are perception, action and language. The focus of the main modality is the human hand. Perceptual aspects of the hand involve touching and manipulation of objects. Much of it is intersensory as in the examples of infant development. Actions of the hand involve learning movement sequences in the acquisition of skills in art and craft. The ability of the hands to construct the conceived objective is an indication of its versatile nature. The hands also convey meaning and language and are vehicles of communication during gestures and signing. Further, the hands are also used in reading text and are able to absorb information in much the same way as the eyes are used for reading. This is supported by the fact that both hands are able as each other in the perception and acquisition of relevant information. Though hand preference is an indication of lateralization, the general ability between the hands is not different.

Although the focus of the manuscript is on the preference and ability of the hands, knowing about the evolution of hand use is important because it is a major contribution to the repertoire of human hand actions. Investigations of limb or foot use in mammals and birds have clarified ways of knowing lateral preference in nonhumans and enriched the understanding of lateralization in humans from comparative investigations. It is of interest therefore to incorporate findings of animal species to know about the evolution of vertebrate lateralization and how this is comparable to lateralization in humans. Of particular interest are the homologies in cerebral hemispheres that may exist between different animal vertebrate species and humans. This may reveal the role of the cerebral hemispheres in the control of actions on either side of the body. It will also enable an understanding of hand or limb use in global or local processing behaviours within particular contexts and the complementary use of the hands to reach targets with one hand/limb while the other hand is used to support the body.

1.9 Evolution of lateralization

There is substantial evidence to show that lateralization is not confined to the human species (Rogers & Andrew 2002). Among humans, the importance of lateralization emerged from observations of functional asymmetries in the brain such as evidence from Broca (1861) on his classic findings of aphasia in Leborne who suffered from speech production following damage to a portion of the left cerebral hemisphere; from Sperry's (1974) split brain patients indicating distinct speech capacities in the left hemisphere and consistent observations of a larger plannum temporal in humans in the left than the right hemisphere (Geshwind & Levistskys 1968). In animals evidence of laterality came easily from observations in lower vertebrates, though an interest in lateralization of function was stimulated by Nottebohm (1971) from his investigations of bird song. Nottebohm found that if the left tracheosyringeal nerve supplying the musculature of the avian syringe was severed, singing was impaired, whereas severing the right nerve had no effect on song production. Subsequent investigations with chickens reported both structural and functional laterality in the avain brain (Rogers & Anson 1979; Andrew, Mench & Rainey 1982).

Observations of motor asymmetries were reported from turning biases or rotations of the body in animals (Bradshaw & Rogers 1993). Rats tend to turn or rotate spontaneously at night in a consistent preferred direction (Glick & Shapiro 1985) and dolphins tend to swim in a counter- clockwise direction when placed in pools (Ridgeway 1986). Mosquito fish have behavioural lateralization both at the individual and population levels and have spontaneous rotational preferences in swimming (Bisazza & Vallortiagara 1996). Individual lateralization refers to characteristics of individual lateral preferences in the population, whereas population lateralization reflects lateral preferences in general such as right handedness in humans. Others have reported right turns in fresh water fish (Goncalves & Hoshino 1990) and tadpoles (Wassersug Naitoh & Yamashita 1999).

Asymmetries in limb usage indicate lateralization in animals. Just as handedness is an obvious asymmetry in humans, avain species that use their foot to manipulate food have shown significant footedness (Friedman & Davis 1938; Rogers 1980; Tomnassi & Vallotiagara 1998). Waters and Denenberg (1994) found right pawedness in a task in rodents (inbred mice) and left pawedness in another reaching task. In primates (Ward et al. 1993) the left hand was used more often to hold food and the right hand for holding on to branches. There are suggestions that as primates became less arboreal, the right hand became available for manipulation and fine motor acts (MacNeilage, Studdert-Kennedy & Lindblom 1987). But handedness has been observed to differ between species. Orangutans for example showed left hand preferences in acts of self grooming such as cleaning teeth, eyes or ears (Rogers & Kaplan 1996) whereas amphibians or frogs showed right pawedness at the population level for removing paper strips from their heads (Bisazza et al. 1996, 1997).

Some species invariant asymmetries have indicated that the different functional specialization of the two sides of the brain is more important than the direction of lateralization. There are some impressive similarities in the direction of lateralization among vertebrates such as the involvement of the right side of the encephalon in spatial tasks for humans (De Renzi 1982). This has also been documented in birds (Rashid & Andrew 1989; Vallortigara, Regolin & Pagni 1999) and in mammals (Crowne et al. 1992; Cowell, Waters & Denenberg 1997). The other cases of species invariance concern the association between the right hemisphere and some species specific behaviours such as escape and attack. The right hemisphere is also involved in aggressive behaviours toward conspecifics and is likely to be evoked when the left eye is used in most vertebrate classes, as in the baboon

(Casperd & Dunbar 1996), chick and adult fowl (Rogers 1991; Rogers et al. 1999), lizard (Deckel 1995) and toad (Robins et al. 1998). Humans usually display intense emotions that are evoked when stimuli are seen with the left eye (Dimond et al. 1976, Gur & Gur 1977). Specialization of the right hemisphere for face recognition in humans (Sergent & Signoret 1992) may have evolved from processes of social recognition in nonhuman species. Split brain monkeys showed specialization of the right hemisphere for face discrimination (Hamilton & Vermeire 1988; Morris & Hopkins 1993). In birds the left eye was involved in the recognition of conspecifics (Vallortigara et al. 2001) and several species of fish showed a left eye bias during scrutiny of conspecifics (Sovrano, Bisazza & Vallotigara 2001). Another example of species invariance is the left hemisphere dominance of vocalization in several different species of birds (Nottebohm 1980; Williams 1990), monkeys (Heffner & Heffner 1986; Petersen et al. 1984), mice (Ehret 1987), frogs (Bauer 1993), and catfish (Fine et al. 1996).

According to some authors, if an individual is lateralized, but the population is not lateralized, the evolutionary significance is debatable. Therefore they prefer to concentrate on lateralization only at the population level (Deneberg 1981). Others maintain that lateralization of function need not be consistent in a population (Rogers 1989) so long as there is functional asymmetry of some kind. Thus in principle lateralization can be considered even if the asymmetries are balanced, so long as sufficient numbers show an asymmetric bias (Collins 1981). However, asymmetries in an individual can arise from fortuitous factors such as illness or an inability to develop complete symmetry. These conditions are also known as fluctuating asymmetries (Leary & Allendof 1989). Not all instances of lateralization at the individual level can be considered as fluctuating asymmetry. There could be advantageous in making a nervous system asymmetric at the individual level such as in saving space for neural computation or evolving separate mechanisms for incompatible purposes (Vallortigara, Rogers & Bisazza 1999). Population lateralization on the other hand may have developed in relation to the need to maintain coordination among individuals in behaviours associated with social life.

1.10 Right and left hemisphere advantages

The earliest evidence for lateralization were from fossil records of primitive chordates (classes of fish) (Jefferies 1979) that seemed to have lain on the right side, thereby receiving different sensory inputs from the left and right sides of the body. Andrews, Tommasi and Ford (2000) suggested that visual detection of prey may have been the first step in the evolution of cerebral lateralization in the ancestors of modern vertebrates. Therefore, lateralization of perceptual functions is ancient and may have occurred just after the brain duplicated itself into two halves with advantages at the individual and the population levels. Advantages at the individual level include enhanced skill in performance and faster responses in lateralized as compared to nonlateralized individuals (Rogers 2000). Rogers reported that lateralized chicks were able to detect a predator with shorter latency than non lateralized chicks. Rogers suggested that having a lateralized brain allows dual attention to the tasks of feeding (right eye, left hemisphere) and vigilance for predators (left eye and right hemisphere). In cats, reaching durations were shorter with their left preferred than the nonpreferred paw and cats that were lateralized had shorter movement times than those that were not lateralized for reaching (Fabre-Thrope et al. 1993). Wild chimpanzees with strong handedness were more efficient at fishing for termites than those with weak handedness (McGrew & Marchant 1999). Therefore early origins of brain lateralization may have been related to simple computational advantages associated with the possession of asymmetric brains by single individuals. The other advantage of lateralization is the need to avoid duplication of function in relation to the saving of neural space (Nottebohm 1977). With time these advantages could be gained and maintained by individuals with asymmetric brains without any need to align the direction of these asymmetries at the population level. However population lateralization evolved so that these behaviourally asymmetric organisms could interact with each other in a homogeneous manner. Consequently the disadvantages of individual asymmetries were protected by the characteristics of lateralized behavior at the population level (Vallortigara & Bisazza 2002).

Studies of lateralization in teleost (bony) fish (e.g. Bisazza, Rogers & Vallortigara 1998; Miklosi & Andrew 1999) have indicated that the features that were found early in vertebrate evolution were the use of the right eye to fixate objects the fish intends to bite and the simultaneous use of the left eye to fixate familiar objects or conspecifics. These features have been found in several species as well as in birds that retain independent eye movements, not having undergone the reorganization of the visual systems that conjugate eye movements have brought in animals (Andrew 2002). According to Andrew, the origin of lateralization occurred as part of the evolution of visually controlled predation. Though both eyes could provide input to the brain to initiate turning toward the prey, only the right eye input was used by structures on the left side of the brain to control mouth reflexes. The right eye has also shown an advantage in recall of food items. Clayton (1993) observed that when marsh tits recalled items that were hoarded with both eyes after a 24 hour period, recall was possible only when they were able to use the right eye. The right eye controlled by the left hemisphere was also able to discriminate between familiar food grain and inedible distracter items (Zappia & Rogers 1983). The chicks were tested monocularly to search for grains of chick-mash scattered on a floor to which small pebbles that roughly matched the food grains in size and hue were adhered. The findings showed that chicks could inhibit pecking at the pebbles and choose grain with their right eye in monocular tests. However when the left hemisphere was treated chemically, the chick's ability to choose grain over pebbles was impaired though the treatment had no effect on the right hemisphere.

Left hemisphere advantages have also been found in mammals such as rodents. Bianki (1988), in a series of experiments, demonstrated left hemisphere advantages in rats for conditioning responses, the discrimination of positively reinforced food stimuli, and maze learning. Bianki argued that the left hemisphere advantages in rats resemble the left hemisphere advantages in humans found for sequential analysis. Subsequently, Bianki (1982) found that the left hemisphere in rats also analyses abstract characteristics of stimuli such as the ability to discriminate unfamiliar geometrical shapes, whereas the right hemisphere could only respond to absolute or concrete characteristics of the stimuli. Bianki (1983) concluded that the hemisphere specializations found in humans (Bradshaw & Rogers 1993) were also found in rats and the left hemisphere is specialized for abstract characteristics of stimuli. This abstract ability of the left hemisphere was also found in chickens (Vallortigara & Andrew 1991) when they were able to differentiate a cage mate from a stranger with their right eye only, and among pigeons (Fersen & Gunturken 1990) that were able to discriminate several abstract visual shapes with their right eye as compared to their left eye.

In chicks the left eye system's advantage for use in topographical information (Rashid & Andrews 1989) may have been the first steps in evolution of the assignment of abilities to maintain orientation within groups of conspecifics. Attacks on conspecifics were more likely with the left eye systems in toads (Vallortigara et al. 1998), lizards (Deckel 1995) and chicks (Rogers 1982). Likewise, in humans, negative responses such as to disturbing events seem to be controlled by the right hemisphere (Hugdahl 1995) indicating phylogenetic similarities in the left hemisphere across species to fear or aggression inducing situations.

Right hemisphere (left eye) advantages in several animal species have shown that the right hemisphere is specialized for spatial processes. Following binocular training in chicks, Rashid & Andrew (1989) found that the left eye allowed chicks to use both distant and local features to guide locomotion, whereas there was an almost complete failure to use distant features with the right eye. The left eye in chicks was also better able to identify the geometrical centre of arenas than their right eye (Tommasi, Vallortigara & Zanforlin 1997) showing that the left eye used position in space better than the right eye. The left eye of chicks also responded to changes in the spatial contexts of the stimulus that were ignored by the right eye (Andrews 1983, 1991). The human evidence for right hemisphere advantages for spatial analysis and global attention (Posner & Petersen 1990) is in agreement with what has been found in chicks. There is evidence that the right hemisphere is specialized for processing spatial information in primates as well. Lacreuse and Fragaszy (1999) found that capuchins showed a stronger left hand preference on haptic and haptic -visual tasks than on simple reaching to targets. Left hand preferences were also found on spatial tasks for gorillas and baboons as they aligned transparent doors to obtain food (Fagot & Vauclair 1988).

Evidence in rodents continued to show that mammalian lateralization resembles that of other vertebrates. Bianki (1988) concluded from several tests that a right hemisphere advantage for spatial analysis exists in both humans and rodents. Right hemisphere performance is better when matching stimuli on arrangements in space, whereas left hemisphere performance is better if space is ignored and matching is based on stimulus properties. Adelstein and Crowne (1991) found that right parietal and not left parietal lesions in rodents impaired the use of allocentric cues during navigation in a water maze. Rats while using the left eye rather than the right eye were able to rely on spatial memory to find the escape platform in a maze (Cowell et al. 1997). All this evidence shows the resemblances between rodents and humans in cerebral lateralization. The left hemisphere free from verbal abilities in rodents is able to categorize objects according to stimulus dimensions, whereas the right hemisphere analyzes stimuli according to their global characteristics.

1.11 Evolution of handedness

Although only humans show a consistent pattern of hand preference in a range of tasks at the population level, evidence of handedness in animals indicated (Mac Neilage, Studdert-Kennedy & Lindblom 1987) distinctions of support and manipulation during hand use. MacNeilage et al. (1987) argued that consistent handedness at a population level could be demonstrated in primates for a particular task. The authors reported that left hand strikes for prey using ballistic reaching movements was evident over 50 million years ago. MacNeilage (1998) attributed these right hemisphere controlled strikes with the left hand as a visuospatiomotor specialization that assisted survival. Tommasi and Vallortigara (1998) observed that support was affected by lateralization in birds. When the eye system in chicks was manipulated by covering one or the other eye, the chicks supported themselves with the foot that was contralateral to the exposed eye. Chicks stood on the foot under the control of the side of the brain that was active for vision. Therefore, task conditions that tend to be controlled by a hemisphere may require that one hand/limb be used to grasp a target and the other be used for support. The tendency for prosimians to use the left hand to catch moving prey (Ward & Hopkins 1993) is a reflection of right hemisphere control for judging spatial positions in the right and left visual hemifields. It is possible that the left hemisphere simultaneously controls the right supporting forelimb so that the left hand is free to strike at the prey. However, evidence from toads (Vallortigara et al. 1998) and birds (Andrew et al. 2000) showed a bias to take food objects present in the right than the left visual field. In this way the right hand may have been in control for taking food. Andrew and Rogers (2002) argued that this could possibly result from the way the food was taken with a medial effector, either the beak in birds or the tongue in frogs which required them to adopt a frontal fixation before striking at the prey, though the decision to strike is made most often when the prey is in the right visual field. Consequently, striking at a prey is likely to occur with the prey in either visual field. Under these circumstances the right hemisphere advantage in attention to both visual hemifields would be likely to put the right hemisphere in control of striking so that the left hand is used to strike prey. Striking with the left paw has also been observed in cats (Lorincz & Fabre-Thrope 1966) and most primates like squirrel monkeys (King & Landau 1993) and prosimians (Ward 1999). MacNeilage et al. (1987) observed the hands of the prosimians showed some specializations such as grasping for support with the right hand and striking prey with the left hand. These acts of grasping branches of trees for support may have strengthened the right hand over time toward a capacity for somatic sensorimotor control, though the preexisting specialization of the left hemisphere for planned movements may have prompted manipulation by the right hand (Andrew & Rogers 2002).

1.12 Advantages of laterlization

Both individual lateralization and population lateralization have their advantages. Individuals that are lateralized in the human population may show skill and efficiency in certain tasks as in left handers that have an advantage for spatial skills over right handers (Gotestam 1990).

Advantages at the population level are mainly the protection of individuals as a group so that there is homogeneity of actions and social predictability. An advantage of lateralization is *increased cognitive capacity* so that the resources of the brain can be channelized for particular processes. Information processing theories (Atkinson & Shiffrin 1968) have indicated that information from sensory modalities move through different stages before an appropriate response is made to simultaneous or discrete presentations of stimuli. Information can also be lateralized so that each hemisphere processes information differently (Andrew & Rogers 2002). Rogers (1982, 1997) observed that chicks hatched from eggs exposed to light were lateralized for attack and feeding responses, but chicks hatched from eggs incubated in the dark were not lateralized for these responses. Rogers (1990) explained that light exposure causes lateralization to develop because the embryo is turned in its egg so that light entering the egg can stimulate its right eye while the left eye is excluded. Lateralization was absent in the dark incubated chicks both at the individual and the population levels, showing reduction in their cognitive capacities of feeding or attack.

Latency for responses is *quicker* in lateralized than nonlaterized chicks indicating vigilance can also be an outcome of lateralization. Rogers (2000) found that chicks exposed to light before hatching were able to detect a predator moving overhead much faster with their left eye than their right eye while feeding, whereas the dark incubated chicks took longer to detect the predator and did not reveal a left eye advantage. Similar results were also observed among cats (Fabre-Thrope et al. 1993) when they showed faster tracking times of a target with their lateralized paw as compared to cats that were not lateralized indicating a functional advantage of lateralization. Lateralization has also shown *efficiency of actions*. Rogers (1997) observed that the chicks exposed to light before they were hatched were better when they were tested monocularly for grain pecking or predator detection whereas in the dark incubated chicks binocular performance was better than monocular performance, indicating that lateralization results in efficiency of use of organs. McGrew and Marchant (1999) observed that wild monkeys with completely lateralized hand preferences were more efficient at fishing for termites than those with incomplete lateralization. In these instances both the forepaws of the monkeys were utilized as in the cat, one for tracking and the other for balance. Similarly, in the chimpanzee one hand held a twig used as a probe and the other served as a stabilizer. Therefore, the complementary use of both hands during task performance was evident across phylogeny.

The advantages of lateralization at the population level are mainly homogenous actions leading to social predictability. Homogenous actions assist the survival of individuals in groups so that the chances of being attacked by a conspecific or a predator are low. According to Rogers (2002), population lateralization is present when there is a need to maintain coordination among individuals in a social context. Evidence showed (Bisazza, Cantaluppo, Capocchiano & Vallortigara 2000) that shoaling in fish was associated with a population bias to turn in one direction when confronted by a model predator. This preserves integrity in the group so that the fish keep apace with each other to avoid a predator collectively. A similar population bias can be observed in any species that moves in a herd or flock (Rogers 2002). There are also indications that a population bias can affect social interactions (Rogers & Workman 1989). When chicks exposed to light before hatching were compared to dark incubated chicks, feeding responses of the two groups showed that the light exposed chicks showed more competition to gain entry to feeding and maintained a hierarchy in the group such that the lowest ranking chick rarely gained entry to the feeding bowl, whereas among the dark incubated chicks the lowest ranking chicks gained access to the food dish more readily, indicating an inability to maintain hierarchy. Thus, advantages of lateralization in populations serve a social purpose and may vary between species or the social context. This account of the evolutionary significance of lateralization enables understanding of how actions of one side of the body emerged to be associated with the control of a particular cerebral hemisphere. It also serves as a preamble to investigations of the development of handedness that is to be discussed in the next chapter.

Examples of hemisphere specialization across phylogeny in fish, birds (chickens), tetrapods (frogs), and mammals (rats, cats, monkeys and primates) consistently revealed the existence of vertebrate lateralization (Andrew & Rogers, 2002). They also show that each cerebral hemisphere is more in control of one type of behavior, like the left eye for identifying conspecifics and the right eye for pecking grain, or the left hand/paw for reaching at targets and the right hand/paw for support in a particular action. Cerebral hemisphere control of the contralateral side of the body has also shown the ability of both sides of the body to engage in purposeful behaviours. Although there are no indications of consistent population level handedness in the animal species, studies on hand preferences in the human population have indicated population level handedness. This does not imply that individual level handedness is not present in the human population. There are also indications that right hand preferences among humans may have emerged as a consequence of gestural communication (Jacquet, Esseily, Rider & Fagard 2012; Esseily, Jacquet & Fagard 2011), and this is different from motor actions of grasping or reaching that may be performed equally well with either hand.

Therefore it is of interest in the present manuscript to review and provide evidence for equal hand ability in performance tasks that do not have a verbal component. Several examples of equal hand ability are available particularly from artisans and skilled performers that use both hands in various occupations such as pottery making, sculpturing, puppetry, or juggling. Furthermore, using one hand as a reference in blind conditions (Millar 2008) while the other pursues a target has shown that each hand is able to successfully pursue a target or serve as a reference. As Brown- Sequard (1847a, 1847b) had long before mentioned, the need for bilateral training in children early in development is necessary to stimulate both hemispheres of the brain so that the brain and spinal cord can develop while all their motor and sensitive elements were exercised. Brown-Sequard also assumed that the stimulation of the minor right hemisphere to equal the potential of the major left hemisphere would produce a superior human being. Therefore, there is sufficient reason to propose a theory of equal hand ability to claim that even if there is a hand preference indicating lateralization at the individual or population level, the ability of the left and the right hands to perform motor actions may not be different. The purpose of the present manuscript is to generate interest in the subject of hand ability in haptic processes so that the reader may appreciate the notion of a theory of equal hand ability.

CHAPTER 2

Hand

The human hand with its multiplicity of functions is a versatile organ and more important, its association with the sense of touch effectively equips the hand for a variety of haptic processes. Sherwood Washburn (1960) affirmed that the modern human brain came into being after the hominid hand became handier with tools, maintaining that the brain was the last organ to evolve. Our knowledge of the ongoing changes in the hand and its effect in the evolution of the brain led to the premise that distinctive human behaviours such as language and tool use are related to brain size. However, when comparing skeletal markers, fossil specimens from arms and hands are rare, than those of skulls, and therefore one- to- one comparison of skull sizes and arms and hands may not be possible (LeMay 1977). Amongst humans, some of the common acts of grasping a hammer or an object involve the thumb and fingers. The grasping action with the thumb, index and middle finger was an early form of hand use.

During the past few decades it was believed that brain lateralization and handedness were only found in humans. Now there are established findings to show that both brain lateralization and handedness at least at the individual level exist in birds (Nottebohm 1970; Rogers & Anson 1979) and nonhuman mammals (rats) (Denenberg et al. 1978). Denenberg suggested that early environmental stimulation like turning chick embryos toward light (Rogers & Anson 1979) or handling rat pups soon after birth by placing them in an enriched environment that contained playthings induced brain laterality for motor behaviours and spatial preferences. These are examples of interactional effects of genetics and early experiences that can induce asymmetries in brain and behavior if the potential for lateralization is present (Cowell & Deneberg 2002).

The corpus callosum is the main bundle of fibers that interconnects the cortical areas of the two cerebral hemispheres (Fix 2002) and an integral part of the neural network that supports higher cognitive functioning (Roessner et al. 2004). In comparative studies Berrebi et al. (1988) using the rat as an animal model to investigate sex differences in the corpus callosum found it to be larger in male than female rats. The authors suggested that sex differences in species should be considered along with effects of environment and development because the direction and degree of asymmetry in rodent behavior is plastic and may be affected by subtle environmental influences. Subsequently Witleson (1989) indicated interactions between hand use behaviours and callosal anatomy. She observed that sex differences in the isthimus of the human corpus callosum varied as a function of right hand consistency. Callosal isthimus was larger in nonconsistent right handed men than in consistent right handed men. This difference was not present in women. Witleson's findings showed that sex differences in callosal anatomy could be an outcome of various interacting factors during development as in the animal model (Berribe et al. 1988) and may be confined to particular cortical regions.

Comparative evidence has shown influences of environmental factors on the establishment of patterns of manual lateralization. An attempt to understand the phylogenetic origins of hand laterality in humans is by examining some of its determinants in nonhumans (e.g. Macneilage et al. 1987; Fagot & Vauclair 1988). In humans, though most models of handedness are explained by genetic determinants (Annett 2002; McManus 1985), the possibility of environmental determinants cannot be ignored (Provins 1997). Damerose and Vauclair (2002) proposed that a primate model of hand lateralization could possibly clarify the origins of human lateralization and its evolution. Some first examples were that of cradling infants in nonhuman primates to know the hand used in this behavior and compare cradling between humans and nonhumans to know the roles of biological and or experiential/cultural factors in the causation of this behavior.

Salk's (1960, 1962, 1973) early studies in cradling infants by both human and non human mothers revealed a left side cradling bias of 80% at the population level in humans regardless of the handedness of the mother and similar cradling biases were observed in nonhuman primates (Manning & Chamberlain 1990). The authors suggested that the left sided cradling would enable the mother to monitor the emotional level of the infant with the left eye controlled by the right hemisphere and this would have a reciprocal effect on the infant that would be able to interpret the emotional condition of the mother. Another motor asymmetry that was studied was head turning, while the infant was lying in a supine position. For both human infants (Coryell & Michel 1978), and nursery reared chimpanzees (Hopkins & Bard 1995) a rightward head turning was observed in most instances.

Grasping or spontaneous reach preferences are likely to predict future handedness of infants. Asymmetries in grasping have been studied to know which hand first grasps a small rattle (Caplan & Kinsbourne 1976). Initially, holding times of the rattle were recorded for one hand at a time before the rattle was dropped followed by holding times for both hands. When a rattle was placed in one hand, it was more often dropped by the right hand, whereas when both hands held rattles, there was no bias between the hands indicating a right hand bias for grasping. Among nonhuman primates there is greater variability in grasp actions. Westergaard, Byrne and Soumi (1998) observed that at 23–24 weeks, tufted capuchin infants showed a significant grasping to the left that disappeared at 47–48 weeks. Fagot and Bard (1995) tested new born chimpanzees for grasping. Grasping responses were observed for the hand and foot while the chimpanzee lay supine on the floor. For both duration and strength of grasping, the right hand and right foot responses were stronger and longer than the left hand or foot.

Damerose and Vauclair (2002) reported the need to involve the mothers as well in infant's preferences and postures of lateral biases. Though children may be influenced by parental attitudes of preference, animals may not be subjected to cultural pressures. Damerose and Vauclair observed lateral preferences in mother olive baboons with their infants. They studied cradling behavior, nipple preference in infants, head position of the infant during cradling or sucking, the hand used for infant retrieval by the mother when the infant is apart from the mother, and maternal carrying when the mother is walking with the infant in one hand while trying to maintain a tripedal position with the other hand. Lateral biases showed that for head and nipple preference there was a right bias to keep the head on the right side and to suckle on the right side. However for maternal cradling, infant retrieval and maternal carrying there were no significant side biases. The results also suggest that it is not clear if there is a relation between maternal cradling, nipple sucking and head position. However, the authors consider it important to have tests that study bimanual actions rather than studying one side preference. Although there is no conclusive evidence concerning the origin of handedness in comparative literature, there are suggestions that it could be influenced through imitation of the mother's hand (Matoba, Masataka & Tanioka 1991), though it is not known how maternal influences can affect hand preference in infants (Hook-Costigan & Rogers 1996).

2.1 Fundamental grips

According to John Napier (1956), there are two major grips of the human hand. The first is the power grip in which the object may be held in a clamp formed by the partly flexed fingers and the palm with counter pressure applied by the thumb, and the second is the precision grip in which the object may be pinched between the flexor aspects of the fingers and the opposing thumb. The thumb and its flexible maneouvers with the wrist allow an extended range of grasping movements. The ulnar opposition is unique to humans and is not found in primates since their thumb is short and relatively inflexible (Marzke 1992, Marzke & Wullstein 1996). Ulnar opposition is the contact between the thumb and the tip of the little finger.

In primates the thumb is short and unable to contact the little finger of the same hand. Opposition of the fingers combined with ulnar deviation of the wrist allows the arm to swing an object like a stick, with force in order to strike. Another effect of the ulnar opposition is the improved precision grip during which small objects are manipulated between the fingers without contacting the palm. The movement of the thumb underlies all the skilled procedures for which the hand is capable. Therefore hand control was found to involve a combination of visual, tactile and proprioceptive feedback on the same action system. With increasing hand control, vision grew to dominate and modulate action. Hence, eye- hand coordination developed and generated theoretical speculations of hand preference. The flexibility of hand actions makes an object external to the body feel as it had sprouted from the hand or foot, and creates a bonding between the hand and the instrument, like fluency in Braille reading or playing a musical instrument. Common examples are those of puppetry and juggling which involve the transformation of a skill into a form of play.

Sir Charles Bell and John Napier initiated the settings for modern studies of the structure and function of the hand. For them the human hand signified a perfect instrument. According to John Napier, a hand without a thumb goes back sixty million years in evolution, when the thumb had no independent movement and was just another digit. Theorists of evolution believe that articulated hand control is an accomplishment of the evolutionary experience of the primate upper limb that has successfully survived processes of selection. However, for humans, it is the task that dictates the grip, and not the tool, as in animals. The task requires a skeletal structure that has a behavioural repertoire with large degrees of freedom in the hand and arm. The hand and the arm in humans have not undergone change in structure with time and have therefore remained functional all through. The Musee de l'Homme at Paris has a fine collection of skeletons of the human forearm dating from the ancient pre historic times to Neolethic and the middle ages, all of which display a commonality in skeletal structure.

The human hand comprising its thumb and fingers has the dual capacity to provide both power and precision actions (Napier 1956). The power grip being the first to evolve is followed by the precision grip. Prehension is the act of grasping and includes processes before and after the grip such as the reach and withdrawal phases of the hand. The many degrees of freedom of the arm and the hand (Napier 1970) in both joint and muscle space contribute to the adaptability of the prehensile act in humans and this functional flexibility is distinct in humans as compared to the relatively limited prehensile actions of primates. Specifically, prehensile movements are those in which the object is held in the palms of the hand and manipulated by the fingers. Nonprehensile acts are those in which the object is not held within the hand, and yet the fingers are in contact with the object, such as in typewriting or piano playing. The skilled use of the hand for prehension and manipulation requires independent finger movements. The intrinsic hand muscles mainly control the configuration of the phalanges of the fingers and abduction/adduction, while the fingers are flexed and extended primarily by the long extrinsic muscles (Landsmeer & Long 1965). Studies on multiple extrinsic and intrinsic hand muscles during gripping have revealed a global spatial strategy in which almost all muscles are activated in some phase of an opposition grasp (Hepp-Reymond et al. 1996). An individual finger movement is not controlled by a single motor neuron, but by different combinations of spatially distributed neurons in the motor cortex (MI) (Schieber 1995). Sensory information from cutaneous receptors in the glabrous (non hairy) skin on the volar surface of the hand and fingers is crucial for efficient hand function. There are about 17,000 cutaneous mechanoreceptors in each hand (Vallbo & Johansson 1984), basically of four types: Merkel, Meissner, Ruffini and Pascini receptors. Each receptor has a specific role and each type has specific characteristics concerning receptive field and adaptation. The importance of cutaneous afferent input for prehension is evident in all forms of actions that involve manipulatory skill and its assistance in the exploratory movements of the hand for haptic information of the object in contact with the fingers.

2.2 Grasping behaviour

Grasping involves the movement of the hand toward the object while shaping it according to the object's size, shape and orientation. Grasping may be defined as the union of the hand and object (MacKenzie & Iberall 1994) and a typical grip pattern describes the anatomical disposition of the fingers and the thumb as they come in contact with the object. All grip patterns are dynamic action events and not static anatomical structures. Reaching and grasping are not initially coupled. During infancy, there is a gradual change as separate movement units become organized into a preprogrammed overall action that takes the hand near the object (Hay 1990; von Hofsten 1991). As the hand moves toward an object, important modifications in the opening-closing sequence of the hand occur. The distance and direction of the reach improve in accuracy at four to five months of age, and the fingers begin to shape according to the size of the object by nine months (von Hofsten & Ronnqvist 1988). During grasping the fingers begin to configure early in the transportation phase, following a closure of the grip till the fingers match the object size.

Early attempts to understand the onset of prehension were mostly naturalistic observations that revealed a lack of consistency in reporting the age at which the hand contacts an object. Moore (1896) reported that casual contact of the hand to an object appears at six weeks of age, whereas Shinn (1893) and Ament (1923) did not observe this behaviour until the tenth week. Grip patterns were seen to have a developmental order, beginning with reflex clasping that progressed to mechanical grasping of objects placed in the hand to voluntary reaching and grasping of objects guided by vision (Shinn 1893). These early reports provide descriptive records of the appearance of prehensile acts, though neither the size nor the form of the object grasped has been described. Besides, reports also indicate that the development of prehension may be dependent on other factors besides vision, such as the constraints of action in varying contexts, while reaching toward or holding objects (Newell & Cesari 1998).

Traditionally, the prehensile act during development was considered to emerge as a reflex and gradually transform into voluntary prehension (Gesell 1928). Maturation was considered to be the regulatory mechanism that stabilized the process of development and the emergence of the fundamental movement patterns including prehension. Development was considered synonymous with growth that followed a stage- wise progression (Gessel & Amatruda 1947) and was directly linked to concomitant neurological changes.

Psychometric attempts to construct tests that assess prehension recorded actions that could be compared to norms of behaviour. Of these, Gessel's (1925, 1928) attempts were the most conclusive and indicate a summary of monthly increments in prehensile behaviour such as the initial development of selective regard for objects from larger to smaller, the development of reaching for objects from bilateral to unilateral and the development of grasping from primary reflexes to grasping in response to tactile stimulation and eventually visually guided grasping.

Experimental studies that followed stressed the role of vision in the development of prehension. For example, Halverson (1931, 1932a, 1932b, 1937) was interested in three forms of behaviour; the nature of visual attention (regard), the manner in which infants reach (approach) and the coordination pattern of the grasp (grip). Halverson observed thumb opposition after 28 weeks and the emergence of grasping with fingertips by 36 weeks. Castner (1932) maintained that grasping a pellet was marked by an increasing dominance and differentiation of the radial digits. In his attempts to study fine prehesion in infants below one year of age, he reported different types of grasping: whole hand closure (32–36 weeks), scissor closure (36–44 weeks) and pincer prehension (52 weeks). Caster's age estimates for fine prehension lag behind that of Gessel (1928) whose test was able to predict the emergence of the pincer grip by 32 weeks of age. These inconsistencies in the onset of finer actions of object contact question the sole importance of visual guidance during development. Still others, for example, McGraw (1945) emphasized the collaboration of visual and motor mechanisms in the development of prehension. McGraw's (1945) emphasis on visual and motor mechanisms listed six phases of voluntary prehension though there was little information about the kind of objects that the infants may have handled. Shirley (1931) stressed the role of vision and noted a three weeks difference between grasping in a lying posture (15 weeks) and the onset of grasping in a sitting posture (18 weeks). Thus these authors emphasized the role of vision, and its importance was considered implicitly or explicitly in the attainment of successful reaching or grasping.

Cognitive accounts of the ontogeny of hand actions attribute the prehensile progression to the development of perceptual – motor integration, with an increasing cognitive capacity and mental representation in older children (Connolly & Elliot 1972; Connolly 1973). They suggest that the sequence of prehensile development is due to the acquisition of prescription of actions known as metaphors, plans or motor programs. That is, children have their own plans or ways of using their hands when dealing with objects. In order to open the lid of a box, the sequence of actions is different from the actions required to open a jar or a bottle. The child demonstrates preparedness that s/he has learnt from past experience with similar objects. Maturational and cognitive viewpoints have been used as alternative explanations though neither is mutually exclusive.

The criteria for grip classification as anatomical (Napier 1956), functional (Landsmeer 1962), or with a dynamic movement component (Elliot & Connolly 1984) termed as 'intrinsic movements' were not very robust and reflected a limitation in the qualitative analysis of arriving at a systematic way of classifying human grip patterns. Newell and Cesari (1998) used a pragmatic system of classifying some basic grips into global functional categories such as (a) picking up an object, (b) receiving an object, (c) projecting an object (as in throwing), and (d) manipulating an object with the fingers. The dynamic approach (Newell & Cesari 1998) in contrast to the maturational and cognitive views argued that prehensile skills were an emergent property arising from the interaction of constraints imposed on the child, and suggested that organisms naturally adopt preferred patterns of coordination that are both stable and adaptive given the constraints imposed on action, such as the goal of the action, object properties and hand size. A particular emphasis of this dynamical orientation in the development of prehension was an understanding of the role of body scale in determining the nature of the emergent grip pattern. The authors proposed dimensionless body/object scaled ratios that correspond to shifts in grip patterns, such as the use of one or two hands for a given set of task constraints which were independent of differences in age. Therefore these authors do not give primary importance to vision in the configuration of grip patterns, but rather emphasize emergent properties of the organism within particular contexts. For example, findings suggested that manipulating the dynamic aspects of the task constraints may result in an earlier emergence of prehension. Von Hofsten (1982) observed that infants as young as five days would reach and make contact with a moving object in the horizontal plane. This indicated that early emergence of the prehensile skill may be determined by the task constraints and may occur before the specific role for visual information in infants reaching and grasping abilities is established (Lockman & Ashmead 1983). Furthermore, the postural position of the infant is also an important task constraint during prehension (Shirley 1931). The infant lying dorsally may be supported in a more stable manner than the one in an upright position. Savelsberg and van der Kamp (1994) observed that the influence of the force of gravity on the development of prehension interacts with postural position. Therefore, prehension may not be dependent on vision, but may emerge spontaneously during interactions with stimuli.

The question as to whether body scale (Newell & Cesari 1998) can dictate the grip in infants and young children remains unanswered so long as they fail to display an awareness of the frictional properties of objects and are unable to anticipate the weight of the object to be lifted (Forssberg, Eliasson, Kinoshita, Westling & Johannson 1995). To attribute the effects of body scaling to infants, such as the use of one or two hands in the lifting of objects would also mean that they possess an internal representation of the object with knowledge of the thumb and finger opposition space. Critics of the dynamic systems approach argue that the dynamic approach might be applicable to only a few domains of development (Aslin 1993) such as motor performance, where many of the components that constrain performance like gravity or limb length can be specified.

Grip patterns usually reflect an understanding of the energy requirements for holding an object. During development, the forces applied to lift an object appropriately decrease prior to the lift to produce smooth lifting movements (Johansson & Westling 1988b; Gordon, Westling, Cole & Johansson 1993). Such anticipatory control of the force output was based on internal representations of the object's physical properties gained during previous experience with the object (Johansson & Westling 1988a). Visual information and knowledge of how useful this information may be for lifting objects may influence the storage and retrieval of information (Gordon, Forssberg, Johansson & Westling 1991). During lifts at which the friction at the digit – object interface was predictable, the object friction slightly influenced the amplitude of the grip force in many children below two years (Forssberg et al. 1995). Young children and adults were required to lift objects weighing 300/200 grams with the preferred thumb and index finger. The younger children usually added a third finger supporting the index finger. The findings indicated a ratio between the grip force on the object and the load force of the object and this ratio was greater among children than adults, particularly for more slippery objects such as silk than sand paper. Therefore children tend to grip the object more forcefully than adults. The slip ratio or the inverse coefficient of friction was higher for silk than sandpaper in both children and adults, though the younger children adopted a greater safety margin with a greater grip force. Alternately, the grip force varied with the texture of the material. After lifting a sandpaper object, the slip ratio for silk was found to decrease, whereas after lifting a silk object, the slip ratio for sandpaper increased. This indicated that the force/lift synergy to be applied for objects was influenced by the representation of forces applied to lift previous objects either similar or different in weight, size or friction. The parallel increase of the grip force and the load force during the loading phase played a key role in grasp stability (Johansson & Westling 1984). These findings do not necessarily suggest that vision is a prerequisite for lifting an object. Rather importance need be given to internal representations of the object that are obtained from feeling the object while lifting, although visual effects such as colour may complement other characteristics such as texture or shape.

Tactile afferent mechanoreceptors in the fingertips mediate adaptation to the object's friction so that the higher grip forces were employed for more slippery objects (Johansson & Westling 1984; Westling & Johansson 1984). This adaptation can be achieved independently at each object-digit interface (Edin, Westling & Johansson 1992). Initial adjustments to new frictional conditions occured just 0.1 second after initial contact (Johansson & Westling 1984). Distinct bursts of slow (SAI) and fast (FAI) adapting afferents occured prior to the force adjustment, with FAI being especially sensitive to the slipperiness of the material (Westling & Johansson 1987). Therefore tactile information was used to adjust the grip force. Children below five years employed more grip force than adults (Forssberg et al. 1995), and only children over two years were able to adapt to the grip forces if there was an unexpected frictional change. The form of the hand in its union with the object emerges from an opposition space (MacKenzie & Iberall 1994). The opposition space provides values for a set of state variables that quantify a posture, such as amount of force to be applied and enervation density for grasping. For example the thumb can serve as an opposition to the forefinger while picking up a needle in order to have the hand adopt a precision posture.

2.3 Development of manual dexterity

Hand function and manual skills have a central role in human ontogeny. Though most hand actions involve a common pattern of activity involving the whole upper limb and trunk, manipulation depends upon the organization and coordination of movements of the digits. Whereas most studies on the development of hand use in young children have focused primarily on the examination of reaching and grasping (e.g. Von Hofsten 1990), two other functional properties of the hand have received relatively less attention: the use of the hand as a perceptual device to explore objects and the manipulation of objects. Manipulation is concerned with the handling of objects to achieve particular goals, and dexterous manipulations refer to being adroit or nimble and performing skillfully. Typically performance has been assessed in terms of speed and accuracy, though little has been known of how tasks are performed or of patterns of digit coordination during skilled actions.

Skilled manual actions consist of a number of components that are intricately related in a given performance. Reaching or moving the hand to a specific location and orientation is coupled spatially and temporarily with grasping (Jeannerod 1997a). Reaching and grasping are not initially coupled. During infancy, there is a gradual change as separate movement units become organized into a preprogrammed overall action that takes the hand near the object (Hay 1990; von Hofsten 1991). Reaching and grasping involve movements of the whole upper limb and often the trunk or the whole body, as in moving to catch a ball. Once grasped, manipulation of the object will depend on the purpose to which it will be put. Moving the hand to lift an object like a book and replace it involves the upper limb and this is known as extrinsic movement since the object is retained in a fixed position within the hand, whereas, if the object is manipulated and retained within the hand, it is known as an intrinsic movement (Elliott & Connolly 1984). Often, two- handed coordination may involve both intrinsic and extrinsic movements as in sewing clothes or brushing hair and the grip force is not affected by the number of finger digits while objects are lifted. Giachritsis and Wing (2010) indicated that though bimanually lifted weights feel lighter than unimanually lifted weights because presumably the perception of increasing number of digits may decrease the total grip force required to lift and hold an object, the total bimanual grip force is not different from unimanual grip force, though perception of lifted weights may seem lighter for bimanual than unimanual lifts.

Regardless of the grip of an object, the more important aspect is the variety of movement that can be afforded in a particular hand task. Arbib, Iberall and Lyons (1995) for example found, that in grasping mugs with different handle sizes, the number of fingers varied while the task remained the same. Therefore when considering grips for hand action such as the precision grip, it is not the accuracy of movement that matters, as the variety of movement (Elloit & Connolly 1984). This increased variety of movement is what affords manipulation. Touwen (1995) found that the ability to use fingers and hands adequately during the manipulation of a prehended object develops earlier and more quickly than adequate anticipatory hand opening during the reaching movement. He attributes this to the capacity of the child to incorporate information about visual size into grasping tasks, which occurs during the toddler period (Gordon, Forssberg, Johannson, Eliasson & Westling 1992). Another aspect of the emergence of digit grips has to do with how the fingers are controlled to take hold of an object. Forssberg et al. (1991) found that the functional synergy coupling grip and load force was not understood in young children. It would require an anticipatory control to assess the object properties and relate them to the grip and load force, and this subsequently emerges during development (Gordon et al. 1992).

Elliot and Connolly (1984) categorized three manipulative behaviours of objects in the hand. (1) Simple synergies that consist of convergent flexor synergies or alternating flexor- extensor synergies, for example, squeezing and releasing a ball or pinching. (2) Reciprocal synergies that involve a combination of movements in which the thumb and the digits show dissimilar or reciprocating movements, such as the flexion of the fingers and the extension of the thumb. The authors refer to it as twiddle or rolling of objects between the thumb and fingers. (3) A sequential pattern that has an independent coordination of the digits in a characteristic sequence, for example, rotary movements as in turning a knob, as the fingers and the wrist rotate in synchrony with the direction of the appropriate movement. Elliot and Connolly (1984) found a gradual increase in the degrees of freedom for movement deployed in action over time indicating increasing choices in hand actions with age. The rigid grips in the younger children change to more flexible grips after the second and third years. The degree of dexterity shown by the hand and digits rely on the corticoneuronal system and its connections (Kuypers 1982).

2.4 Role of vision in hand actions

The precision grip with its opposition of the thumb and forefinger is considered to be the hallmark of manual dexterity. The ability to grasp with the precision grip is not present when voluntary reaching and grasping emerge. In humans the precision grip appears around 9 to 12 months after birth and grows in strength and diversity with increasing corticoneuronal connections (Kuypers 1982). During reaching, the configuration of the hand begins long before the object is contacted. Visuo-motor channels are thought to possess extrinsic object properties such as position and orientation in extrapersonal space and intrinsic object properties such as the size and shape required to regulate transport and grasp effectively (Jeannerod 1981, 1984). The potential impact of anatomic and task constraints in prehensile skills reflect a cognitive capacity in the development of prehension (Connolly 1970). That is, as children get older they are more able to hold or handle larger or complex objects. In contrast, Newell and Cesari (1998) claim that prehensile skills are affected by relating hand to object size, and emphasizes the dynamics of emergent properties as a function of the organism, the environment and the task. They found that grip size was influenced by the geometry of the object and hand size and follows a body scaling that is independent of age.

Theories of eye- hand coordination have long emphasized that hand preferences are an outcome of visuo-motor control (e.g. Bell 1840; Woodworth 1898). Experimental findings of reaching in infants have supported the idea, visual guidance of the hand is necessary for successful object contact (e.g. Halverson 1937; Caster 1932; Shirley 1931; McGraw 1945). Subsequent experiments as well (White 1970 and White, Castle & Held 1964) stressed the development of visual motor capacities and its relation to visually directed reaching. White and Held (1966) also indicated that prehension to visual stimuli can be accelerated by modifying and enriching the conditions in which the child is reared. Others (Bower 1974) observed that visually adapted reaching and pre- reaching in young infants begin at one to two weeks as infants adapt their reaching to the relative size and distance of objects and object direction (von Hofsten 1982). Thus these authors consider visual guidance to be necessary for the acquisition of hand actions and attribute the importance of vision implicitly or explicitly to the attainment of successful reaching or grasping.

2.5 Experimental treatments that assess the role of vision

2.5.1 Restricted rearing

The most direct way of assessing the effects of visual guidance in reaching an object is to deprive the individual the opportunity to use the sense modality of vision from birth. The individual could then be tested for perceptual capabilities in the deprived modality during development as well as adulthood. If the perceptual capacities are poorly developed as a consequence of visual deprivation as compared to sighted people (so long as the individual does not suffer from any mental deficit or neurological injury), it would demonstrate the role of visual experience in learning and development. This technique is known as restricted rearing, though restricted rearing conditions are considered to be unethical if conducted with human observers because of the possibility of producing long- lasting perceptual difficulties. However, certain clinical conditions that reproduce the circumstances needed to study the effects of experience on perception have been valuable. For example, Von Sendon (1960) reported cases of cataract that regained vision later in life. After the removal of cataract, these individuals were unable to recognize objects by sight, though they were able to recognize the same object by touch. For instance when asked whether an object was a square or a circle, these individuals had to undertake the painstaking procedure of counting the corners of the figure before the forms could be distinguished. They were only able to distinguish the presence or absence of an object in their visual field, but were unable to identify the object visually. This indicates that in the absence of training with the visual modality, individuals rely on other sources of information in transacting with the environment. Similarly Gregory and Wallace (1963) reported a case of congenital cataract that received sight much later in life after surgery. When the patient was required to look and tell time from a wrist watch, it was observed that he was unable to tell time till he touched the small and large needles of the watch to know their positions. It seemed to him that he was touching with his eyes. These findings also indicate that vision may be a complementary source of information rather than an essential source of information in instances when the content is not modality specific.

When restricted rearing studies were conducted with animals, a common procedure was to dark rear the animal from birth. It has been observed that successfully dark reared animals tend to show perceptual deficits. They have difficulty in discriminating visual patterns and show erratic or abnormal responses to the depth or distance of targets (Reisen & Aarons 1959). The relationship between perceptual development and visual experience has been found to be complex. Fantz (1965) found that there may be a critical period during which an animal will be able to respond to visual stimuli if it is to see normally. Animals that were dark reared for a month's time were able to recover their perceptual abilities after a few days of exposure to light, whereas those that were dark reared for longer periods were unable to reach normal levels of functioning. Therefore the onset of blindness or loss of vision may affect the integration of information and this was evident in conditions of a task specific nature that may require vision.

2.5.2 Reafference studies

Along with vision, movement of the entire body is considered to be important for perceptual development. Stimulus input as a result of an individual's own movement is known as reafference, whereas stimulus input that acts on a passive observer is known as exafference (Von Holst & Mittelsteadt 1950). In a series of classic experiments Held and his associates indicated the relevance of reafference in visually guided spatial behaviour. Held and Hein (1963) dark reared kittens till the age of about twelve weeks, after which they received periodic patterned visual exposure in two ways. One of the animals leashed to a rotating rod, was allowed to walk around the compartment, while the other was moved about in a basket that was yoked to the other end of the rotating rod. Both animals moved at the same speed and direction. The moving kitten experienced changing visual stimulation as a result of its own movements, while the passively moved kitten experienced the same stimulation, though not as a result of its self- generated movements. Later when both animals were tested for depth perception, the active animal was able to dodge a rapidly approaching object, as well as place his paw appropriately to avoid collision, whereas the passively moved kitten was unable to perform these tasks. These findings indicated that movement in space was necessary to develop coordination between vision and action. Hein, Held and Gower (1970) repeated this procedure with the pair of kittens while exposing both kittens to active and passive movement treatments. One eye received active exposure, while the other eye received passive exposure. When tested with the actively exposed eye, the kittens seemed to have depth perception, though they failed the tests when tested with the passively exposed eye. This indicated that experiential effects were specific in depth perception, and suggested the need for visual guidance to locate objects in space. The findings also suggested that visual guidance of the hand or the paw was necessary for reaching toward objects. The simple actions of reaching an object would require not only information of the distance and size of the object to be reached, but also the ability to guide the limb on the basis of the perceptual information. In instances when the paws of animals had been occluded by collars (Held & Hein 1967) in a lightened compartment after the kittens were dark reared, it was observed that the ability of the kittens to place their paws by visually directing them toward targets was quite poor. Further, Hein and Diamond (1971) found that the specific exposure of the paw was necessary for reaching a target accurately. They covered one paw of the kitten with an opaque material and another with transparent material. When tested later for paw placing, the kittens were only able to place their exposed paw on the targets. The need for visually exposing limbs for accurate reaching was also demonstrated with a monkey by Held and Bauer (1967). The monkey was reared without exposure of its arm for 34 days and later allowed to see its limb. The monkey displayed surprise at seeing its arm and was unable to accurately extend it for reaching objects. White (1971) reported instances of human infants who spent a good deal of their time during the first month of their life watching their hands. Though their reaching was quite inaccurate initially, it improved steadily with visual guidance showing vision provides feedback for movement accuracy.

2.5.3 Perceptual rearrangement

Studies of perceptual rearrangement have shown that when the context is altered temporarily by wearing prisms, subjects were able to adapt themselves to the altered conditions after a period of practice. These aftereffects indicated that subjects learn to adapt to the rearrangements by correcting the movement of their hand or body in space with guidance from vision. Researchers have attempted to explain the conditions for adaptation to rearranged stimulation. Held and Hein (1958) argued that adaptation depends upon active movements, as does the development of visually guided behaviour. Observers that were permitted to actively move their arm in the rearranged condition were better able to adapt to the distortion produced by the prism than observers who had their arm passively moved by the experimenter. The role of movement in adapting to rearranged environments was also evident among observers who were allowed to walk about the room while wearing prism goggles as against those who were wheeled around in a chair while wearing the goggles (Mikaelian & Held 1964). An important aspect about active movement is that it provides error feedback, which informs the observer about the direction and extent of the movement (Howard, Craske & Templeton 1965). It has been argued that vision provides relevant information during adaptation. Adaptation is not only indicated by some proprioceptive component as a change in the felt positions of different parts of the body, but adaptation to visual displacements have been observed in animals when the nerves that provide information about the position of the arm had been severed (Bossom & Ommaya 1968; Taub & Berman 1968). Foley (1970, 1974) demonstrated the importance of vision in adaptation by distorting the direction of displacement in each eye. Either, one eye of the animal had an upward displacement and the other eye had a downward displacement, or one eye adapted to the left side and the other adapted to the right side of the visual field. After several hours of exposure it was found that each eye adapted to its own particular distortion. It seemed that adaptation to optically rearranged stimuli involves a kind of perceptual learning that alters the appearance of visual space.

Recent evidence indicates that adaptation to eye hand coordination may get less stable as an individual advances in age. Guan and Wade (2000) tested younger and older adults for pointing at straight ahead visual targets while looking through laterally displacing prisms, with the hand visible early in the pointing movement. Aftereffect tests indicated that the eye hand shift was smaller for the older subjects, whereas the proprioceptive shift did not differ between the older and the younger groups. The authors concluded that perceptual motor adaptability declines with advancing age. These findings question the stability of the visual system over time and suggest that the learning of visually guided behaviour in reaching for targets may not be always sustained. An early finding (Hamilton 1967) with split brain monkeys under normal and prism altered eye hand coordination indicated that the effects of adaptation to wedge prisms were associated with changes in the proprioceptive-motor mechanisms of the exposed limb rather than with visual mechanisms.

2.6 Eye- hand coordination

Theories of eye hand coordination gained attention since Sir Charles Bell in his treatise on the 'hand' wrote that no serious account of human life can ignore the central importance of the hand. He asserted that both the hand and the eye develop as sense organs through practice, which would mean that the brain teaches itself to synthesize visual and tactile perceptions by making the hand and eye learn to work together. According to Bell the learning process must involve the correlation of sensory information from retinal (light) and cutaneous (tactile) receptors with a muscle sense. The brain would actively orient the receptors in the eye or the hand toward the object of interest and move them during the process of exploration. The representations in the brain would be based on the information from the movements of the eye and limb, retinal or skin receptors.

Studies of eye hand coordination have been increasingly supported by subsequent evidence that the upright position of the head in young infants between five and eight weeks of age increases their reaching accuracy to match that of infants that are twenty weeks of age (Jeannerod 1997a). Painstaking research with infants has shown that eye hand coordination can be observed very early in development. Von Hofsten (1982) observed that the arm-hand movements of neonates to slowly moving objects were aimed closer to the object than other movements, while the neonate fixated the object. This effect was not a function of body posture or head direction accompanying changes in visual direction. Fixated movements clustered closer around the object and the hand was found to slow down near the object in the best direction of the fixated movements. Von Hofsten concluded that there exists in the newborn a rudimentary form of eye- hand coordination which is primarily attention arousing than manipulative. Others have argued that reaching in the first half of life can be distinguished from reflex acts (McDonnell 1979) and that there is sufficient evidence that hand movements of infants under 8 weeks of age are progressively coordinated with visual stimuli. Furthermore, up to at least 12 weeks of age, the left hand showed greater activity and orientation to visual stimuli than the right hand, which was contralateral to the dominance of neonatal reflexes. The emergence of visually guided reaching and development during the second half of the first year was observed to be nonreflexive and instrumental. Subsequent findings indicated that infants do not reach to objects according to shape till they are able to distinguish the characteristics of the objects (Pieraut-le -Bonniec 1985). Infants aged 3 to 11 months were tested for visuomotor coordination of concave, convex and flat objects. The younger infants did not adjust the shape of their hands prior to touching stimulus surfaces. Although by ten months all the infants modified the shape of their hands in accordance to the shape of the object prior to contacting the object, the establishment of coordination between visual and haptic perception enabled haptic perception to supply vision with significant information. Infants relied on the haptic information of shape to configure their hands to reach an object.

Other studies indicated that visual gaze and eye orientation toward targets enable reaching or pointing. Enright (1995) demonstrated that eye orientation at the time of pointing has a systematic influence on a delayed pointing response to the remembered location. Remembered locations as indicated in total darkness were more precisely reproduced when the subject's eyes were aimed toward the target before pointing, than when the eye positions were straight ahead while pointing, thus demonstrating the nonvisual impact of eye orientation on eye-hand coordination for the spatial memory of targets. Others (Johansson, Westling, Baekstrom & Flanagan 2001) have shown the relevance of visual gaze supporting hand movement planning to reach a target, by marking key positions to which the fingertips were directed to the object to be grasped. The authors concluded that the salience of gaze targets arises from the functional sensorimotor requirements of the task. Soechting, Engel and Flanders (2001) found a relation between pointing to remembered targets and visual gaze and suggested a model of limb control in which gaze position can provide the target signal for limb movement.

Theorists have attributed the accuracy of reaching and object perception to the neural control of specific visual pathways (Jeannerod 1994, 1997a). They argue that visual pathways mediating prehension indicate distinct pathways for visual perception and visual control of movements (Goodale & Milner 1992), and offer suggestions for different visual pathways in the dorsal stream itself which corresponds to the hypothesis of various visuo-motor channels (Jeannerod 1994), one for reaching and another for grasping. Jeannerod (1984) observed that the maximum grip size occurs during the first 60–70% of the reach, before the fingers come in contact with the object, and the velocity profile of the transport component indicates that the entire movement is programmed in advance (Brooks 1984). Marteniuk, Leavitt, MacKenzie and Athenes (1990) found that, for an increase of 1cm in object size, the maximum grip size increases by .77cm, indicating that the rate of grip aperture can be used as an index of the functioning of the visual system responsible for immediate processing. According to Jeannerod (1997b) the hand is at the interface between visual and tactile analysis and its movements are initially dictated by vision. It is only after contact with the object that the hand becomes a tactile organ. Grasping an object is not only an indication of manual ability, but also reflects visuomotor transformation or the process by which visual attributes of an object are translated into motor commands (Jeannerod 1997b). Jeannerod and Decety (1990) demonstrated that the grip of the object indicated by the distance between the thumb and the index finger matched the size of objects presented through a mirror precluding vision of the hand. Their results showed that the visuomotor estimation of size is linearly related to target size and therefore reflects the objective size of the target, unlike instances in which subjective scaling of physical size produces a power function with an exponent and therefore reflects apparent size (Teghtsoonian 1965; Stevens & Guirao 1963). Thus according to Jeannerod, vision seems imperative for successfully reaching an object.

Neurophysiological findings with monkeys (Battaglia et al. 2001) for visual fixation, reaching, and saccadic tasks showed that reach- related cells were influenced by eye position information. The activity of most neurons related to a combination of both hand and eye signals and revealed the existence of a selective match between activity types and parieto- frontal connections. It was concluded that there is a predominance of eye- related and preparatory information in the rostral part of the premotor cortex, and a preponderance of hand related information in the caudal part. Carey (2000) reviewed evidence that suggested eye movements may play a more direct control of hand actions after targets have been identified. Single unit recordings revealed both hand and eye movement related activity in the parietal cortex of the macaque monkey. Experiments also demonstrated yoking of eye and hand movement and eye- hand symbiosis in humans.

Reviews continue to show (Buneo & Anderson 2006) that the posterior-parietal cortex plays a role in converting sensory information to motor commands. The posterior- parietal cortex is also involved in integrating sensory input with previous and ongoing motor commands to maintain a continuous estimate of the state of the arm for updating movement plans. These actions are encoded in an eye- centred frame of reference. A number of behavioural studies in humans and monkeys that have investigated reach related areas have emphasized the primacy of eye- centred coordinates in movement planning. The stimulation of the superior colliculus in monkeys suggested that this structure encodes gaze direction in retinal coordinates (Kleir, Wang & Crawford 2001). Population of arm movement related neurons that are eye centred have been identified in the superior colliculus (Stuphorn, Bauswein & Hoffman 2000) the ventral premotor cortex (Schwartz, Moran & Reina 2004), and the dorsal premotor cortex (Shen & Alexander 1997b). In some instances these neurons were found to coexist with ones that seemed arm- or hand- centred.

Some authors consider vision to have a profound effect on arm movement very soon after birth (Vandermeer, Vanderweel & Lee 1995). The strong influence of vision on arm movements was related to encoding targets to reach in eye- centred coordinates demonstrating the role of the posterior- parietal cortex in the updating of reach planning (Mendendorp, Goltz, Vilis & Crawford 2003). Examples of patients with optic ataxia (Khan et al. 2005) an inability to accurately reach targets, revealed a disorder associated with damage to the posterior parietal cortex. The patient's behavior showed that the reaching errors that characterized this disorder depended on the perceived position of the reach target in eye coordinates at the time of reaching. There are suggestions that for target locations with some visual feedback of the moving limb, the variable error (McIntyre, Stratta & Lacquaniti 1998) and constant error (Flanders, Helms-Tillery, Soechting et al. 1992) patterns of the reaches suggest a coordinate system centred on the line of sight, whereas for movements of remembered locations made in the dark, constant error patterns suggested a coordinate system centred on the shoulder of the pointing arm (Flanders et al. 1992) and variable errors were an outcome of the starting position of the effector with respect to the body (McIntyre et al. 1998). Undoubtedly, vision provides complementary information by increasing the accuracy of reaching and grasping. However, the possibility of prehension and manual dexterity in the absence of vision needs be discussed before attributing reaching or grasping solely to the effects of eye hand coordination.

2.7 Reaching without vision

The predominance of studies in reaching and grasping attribute successful performance to the ability of the visual modality for aligning the hand and arm to the size and orientation of the object (Halverson 1931, 1932a, 1932b, 1937; Caster 1932; McGraw 1945; Bower 1974; von Hofsten 1982; Jeannerod 1994, 1997a, b). The role of vision is undoubtedly facilitating, but speculations as to whether it is a necessary modality in the attainment of these specific behaviours can only be examined by comparing congenitally blind children with sighted cohorts. Fraiberg's (1968) extensive work with eight blind infants showed that these infants first reached for objects only at the age of ten months, whereas sighted infants on an average reached at the age of five months. Fraiberg (1968) observed that developmental delays in other behaviours of locomotion such as crawling suggested that reaching is the critical skill to locomotion. According to Fraiberg reaching in blind infants is a two stage process where the initial reaches are to sounding objects pulled from the children's hands and this is followed by reaching to sounding objects held directly before them. Whereas sighted infants spontaneously reach for objects they see, blind infants need to be prompted by sound. If sounding objects are not present, the blind infant may not reach. Therefore reaching is dependent on the awareness of spatial information about the object. Subsequently Adelson and Fraiberg (1974) and Fraiberg (1977), compared the ages of achievement at different motor abilities, in a longitudinal study of ten congenitally blind children with the norms of sighted children by Baley (1969). Though there were no differences with the sighted children in aspects concerning postural control, such as sitting down momentarily, rolling from back to stomach, sitting steadily, walking supported by hands, or standing alone, blind children showed a delay in motor abilities requiring self- initiated mobility such as elevating themselves using their arms, raising themselves to sitting position, raising themselves to a standing position either with the support of furniture, or walking alone and crawling. Therefore, those behaviours requiring postural control were unaffected, whereas those requiring self-initiation and mobility were delayed in blind children. Sighted children reach or move toward objects they see, whereas blind children need to be told about toys or furniture about them. Fraiberg associated these delays with lack of incentives for locomotion. For example, Friedman (1971) observed sighted socially deprived infants as crawling on a par with blind infants, indicating that the deficit is of a non-modality specific kind, including not only the lack of visual but also proprioceptive, kinesthetic and auditory stimulation.

Fraiberg (1977) in her in- depth study of the evolution of grasping behaviour in ten blind children observed that at three months of age, as many blind children as sighted children had their hands opened; however at five months of age the blind children did not grasp their own hands nor play with their fingers as did the sighted children of the same age. They were unable to hold two cubes in one hand and interchange them between the hands, as sighted children normally do. Also, the blind children at nine months did not use pincer grasping which involves the opposition of the thumb with the forefinger, but instead used palmer grasping. Pincer grasping, though found to appear only by ten months of age, was scarcely used (Fraiberg 1968). Therefore in the absence of vision, some behaviours in the attainment of perceptual motor skills such as manipulating objects with the fingers may seem delayed. Indeed, Hatwell (1987) has cogently argued that in instances when blind infants do not use their hands as a perceptual information seeking device (e.g. Fraigberg 1977), the reaching behaviour of these infants is mainly a motor executive one where, for example, the hand is usually used for putting objects in the mouth, and the deficit if any is non-modality specific (e.g. Friedman 1971). When blind infants reach, this behaviour may be related to their conceptual development (Bigelow 1986). For example, the ear- hand coordination for reaching objects in blind children is attained at eight months, whereas eye hand coordination in sighted children is attained by four months. Nevertheless sighted children are not able to reach a hidden object they hear until they are about eight or nine months of age and this is at par with the ages of the attainment of object permanence in blind children. Therefore blind and sighted children do not differ in their understanding of the object concept and vision is a necessary modality for neither the attainment of object permanence nor subsequent cognitive development.

Bigelow (1986) carefully studied the development of reaching in blind infants, with toys that could be continuously touched or heard. She investigated whether touch cues or sound cues were more important for reaching in the absence of vision and if cues would facilitate reaching. Bigelow (1986) found that in instances in which touch and sound were used in analogous tasks, the touch tasks were easier. The infant would rather reach for a toy placed on its chest than for a rattle held before her/him. Children responded to continuous touch before they responded to continuous sound and children reached to cues of previous touch before they reached to cues of previous sound. When touch and sound cues were in conflict, children initially respond to touch cues. Fraiberg (1968, 1977) observed that blind children reached at the body mid-line such as the chest for sounding objects taken from their hands before they reached at mid-line for objects on sound cues alone. Fraiberg concluded that the mid-line was the first space to have subjective reality for blind infants.

There is some evidence that Fraiberg's findings of reaching in blind infants may be operational in sighted infants as well. Wishart, Bower and Dunkheld (1978) studied reaching in sighted infants in dark conditions, so that though infants heard the object they were unable to see the object, nor know the direction of their reach. They found that early reaches at four or five months followed a sharp decline at ten months of age. The authors reasoned that the early reaches were an outcome of intersensory integration, and the later ones revealed the existence of objects. Wishart et al. (1978) found that the later reaches increased when the infants accidentally came in contact with the sounding object suggesting that, like Fraiberg's and Bigelow's findings with blind infants, prior contact with the physical objects facilitated reaches in sighted infants as well. Furthermore, Wishart et al. (1978) observed that, as in the case of blind infants, directional reaches to objects as in off centre positions were more difficult than reaches to objects at the mid-line for sighted infants. Therefore, there are no fundamental differences in reaching with or without vision. Blind children attain comparable performance with the sighted by relying on self-referent cues (Millar 1981b) and self referent cues are reliable (Stelmach & Larish 1980). Fraiberg (1968) has shown that this is evident in early infancy when infants reach to their body mid-line for objects they have not seen. Thus the difference between the blind and sighted children may be in their strategies for attaining information and is not a function of visuo-motor control (Jeannerod 1984, 1994). The fact that blind children between the ages of 6 and 15 years were able to perform acts of reaching, grasping and assessing objects of various sizes using the precision grip (Ittyerah 1993, 2000) separately with the preferred and nonpreferred hands is an indication that though vision may provide complementary information, convergent information is attained from other sources (Millar 1994) and therefore vision is not a necessary condition for tactile hand ability tasks.

Hatwell (1987) has argued that early in development infants rely on tactile information and perceive objects held in their hands without much assistance from vision. It is only after the age of five or six months that vision dominates and infants become dependent on eye hand coordination. For example, when the hands of the infants were occluded by a screen (Streri & Pecheux 1986a, 1986b), so that they could not see objects held in their hands, infants displayed a haptic habituation to the familiar stimulus and a novelty response to a new shape as early as four months of age. After five months, there seemed to be an increased synchronization between manual and visual inspection of objects (Rochat 1985). Hatwell (1987) reported that infants aged five to six months displayed suppression in their grasping response when a screen was placed on their shoulders to prevent them from seeing their hands. Although these children did not display any signs of distress nor try to remove the screen, they did not close their hands on the object the experimenter tried to put in it and their hands tended to avoid contact with the object. Over 60% of the infants failed to perform similar unimodal haptic tasks after five months of age or above whereas only 20% failed to perform unimodal haptic tasks at ages less than four months. Therefore vision clearly dominates at older ages and infant's haptic perceptual abilities begin to be underutilized. Thus early in development, the tactile- haptic system takes precedence over visuo-motor channels for object perception and action.

2.8 Is pointing guided by vision?

Pointing in human infants follows the onset of prehension and involves processes such as reaching and hand withdrawal. Pointing is also an early indication of communication and socialization. Hand pointing responses are a valuable measure of performance in spatial localization. This has been little studied till recently, though more often, target position encoding by pointing in the absence of vision has been used to know spatial localization in normal people and patients (Rossetti 1998; Paillard 1991). Pointing in human infants emerges toward the end of the first year but continues to develop until the age of 2 years when infants use pointing to direct the attention of the mother to objects in space (Bates et al. 1994). Pointing is thus a critical behavior to ascertain the understanding of self in relation to the physical environment. Clarifications about this behavior are given by studying underlying spatial representations. Indeed, pointing at targets is an easy and simple way to objectify divergent types of spatial representations and the underlying mechanisms (Paillard 1991).

Pointing requires the control of precise hand movements in space. These movements are controlled by vision in two ways. First vision is thought to provide information about the position and movements of the hand and then guide the hand to the target. A number of studies suggest that pointing accuracy deteriorates without vision (Elliot, Calvert, Jaeger & Jones 1990; Elliott, Carson, Goodman & Chua 1991). It has been observed that memory for target location decays quickly following visual occlusion. While pointing to visual targets with the dominant hand, accurate location information is retained for up to 2 seconds and then rapidly decays (Elliot & Madalena 1987). In dark conditions subjects were unable to place their index finger on the target spot of light because they were not able to see the pointing hand (Buyakas, Vardanyan & Gippenreiter 1980). Accuracy for placing pins at positions underneath a table top seen or felt on the table was best under visually informed conditions in children at ages of 4 to 12 years (Von Hofsten & Rosblad 1988). Undoubtedly, the availability of vision or vision and proprioception would enhance performance as the modalities provide convergent information (Millar 1971, 1994), whereas proprioception by itself in the absence of vision would be less facilitating. Von Hofsten and Rosblad claimed that the visual condition reduced both random and systematic errors, though in the memory condition when the child had to look at the target dot, memorize the position, close the eyes and place the pin underneath the table top on which the dot was placed, there was more error, than in the condition that did not provide any visual information. Memory for vision was not better than that for touch in this instance. This raised the question as to what is remembered in pointing tasks. Rosenbaum, Meulenbrock and Vaughan (1999) reported that subjects remember the final postures of their fingers in pointing tasks, and not the final spatial locations that may be visually guided. Rosenbaum et al. claim that stored postures form the basis for movement planning. These authors argued that if people store and use postures, then they should copy the posture adopted with one arm to the other arm when possible. Their findings indicated that final postures were remembered and were copied from one arm to the other when subjects tried to reach repeatedly to the same location in the midsagittal plane with alternating arms or the same arm. Memory for postures was a reflection of the memory for final body states and this was not solely determined by vision.

Furthermore, studies have consistently reported that processes underlying pointing are early and that infants do not need to see their hand to reach and contact a toy. For instance infants differentially oriented their hand to a similar extent regardless of lighting condition (McCarty, Clifton, Ashmead, Lee & Goubet 2001). Similarly, babies between 6 and 25 weeks of age did not rely on vision of the hand when attempting to reach for an object (Clifton, Muir, Ashmead & Clarkson 1993). Babies contacted glowing objects in the dark, when vision of the hand was restricted, at the same rate as during normal daylight conditions, when they were able to see their hands. In an earlier study, Trevarthen (1984) tested infants

16 to 20 weeks of age, that is, just at an age when infants begin to show their first goal-directed reaches. A lightweight screen attached with a headband to the baby occluded the view of arm and hand in either central or peripheral vision. None of the infants attempted to see their arm or hand before initiating a reach or while executing the movement. These empirical findings indicate that proprioception enables accuracy for reaching, and visual guidance of the hand is not necessary to reach an object.

Subsequently, studies have indicated (Lynch, Lee, Bhat & Gallaway 2008) that infants do not have any stable arm preferences during the prereaching period. They examined left and right hand kinematics from 13 healthy infants during trials with and without a toy present from 8 weeks of age through the week of reach onset. When a toy was not present, the right hand moved faster, yet ended further away from the midline. With a toy present, the right hand moved longer lengths, yet ended movements further away from the toy. The authors observed that although infants begin adapting their prereaching kinematics many weeks before reach onset, there was no evidence of a systematic right-left difference before reach onset in hand movements with or without the presence of a toy.

Studies with sighted infants have demonstrated early right hand preferences for discriminating features of objects and left hand preferences for contours of objects (Streri 2002). Streri (2005) observed that at two months of age when prehension–vision coordination was absent, the tactile sense was predominantly utilized and touch- to- vision transfer of information was observed. At five months when reaching and grasping an object was intensive, vision dominated showing successful vision- to- touch transfer though little touch- to- vision transfer. Although Streri's tasks involved the manipulation of three dimensional shapes, they resembled pointing and reaching tasks in structure. Furthermore, among blind children that have had no exposure to vision during development, there was a possibility that they relied on convergent information from modalities other than vision (Streri 2005). Therefore the hands are well utilized for perceptual and motor functions early in development. Ittyerah, Gaunet & Rossetti (2007) found that blind children occupied lesser surface area of pointing to targets on a touch screen as compared to the sighted children, showing better accuracy of pointing at the targets. This finding revealed a steady tactile development in the blind during infancy (Streri 2005), and suggested the development of some parts of the brain that may undergo significant postnatal development.

The findings of Ittyerah et al. (2007) indicate that vision is not necessary for a spatial framework since consistent effects of development have been found in blind children as in their blindfolded sighted peers. The findings suggest that congenital blindness may not be detrimental to the development of spatial representations, since there is evidence that people who are totally blind from birth can perform as well (Millar 1994) or more proficiently (Hollins 1986) than the sighted on spatial tasks. Casteillo, Bennett, and Mucignat (1993) for example observed that experience of vision was not necessary for the coordination or patterning of the basic reach to grasp movement in blind adults. A recent case study provides a possible explanation for the lack of performance differences between the blindfolded sighted and congenitally blind children. A blind born lady (Ostrovsky, Andalman, & Sinha 2006) who underwent surgery for the removal of dense congenital cataracts at the age of twelve years revealed that she exhibited a high level of proficiency on most form and face perception tests twenty years later, with a visual acuity of 20/200. While this finding does not rule out residual impairments, it suggests that significant functional recovery is possible even after several years of congenital visual deprivation. Recent evidence (Millar & Al-Attar 2005) indicates that vision improves performance in a haptic spatial task only in so far as it adds cues that are potentially relevant to spatial discrimination and reference. Vision does not affect haptic processing if it does not add task- relevant information.

2.9 Hand preference

If reaching and pointing actions can be performed in the absence of vision, is vision a necessary prerequisite for the choice of the preferred hand?

When one hand is preferred over the other for various tasks such as writing or sewing, one can expect an enhancement in the ability of that hand over the other. It is also an indication of the choice of the preferred hand and is referred to as handedness, and quite often reveals the skill of the preferred hand. Handedness is a measure of performance or output processing and is different from hand advantage. Most people have been found to be right handed by traditional hand preference questionnaires (Annett 1985). However, a left hand advantage in certain tasks for speed or accuracy of recognition does not indicate that the person is left handed. Hand advantages, as for the left hand in tactual recognition, need to be distinguished from the notion of handedness. Hand advantages are considered to be controlled by the contralateral cerebral hemisphere and display input processing, for example, of non-verbal spatial tasks (Kumar 1977; Dodds 1978), or right hemisphere superiority for tactual recognition. Instances of hand advantage for tactual recognition have indicated that the preferred hand may not always be the skilled hand. For example, Witelson (1974) had children palpate two nonsense forms one with each hand simultaneously. Subjects then selected the two forms that they were required to palpate from six visually presented stimuli. Boys from grades 1 to 8 showed a left hand advantage for the forms and a right hand advantage for pairs of letters they had felt similarly. However not all tactual performance depends on systems which are entirely lateralized as in instances of discrimination of roughness, texture and passive tactual discrimination (Gazzaniga & Ledoux 1978).

Hand ability refers to the potential ability of either hand and is considered to be uniform for both hands, unlike hand preference or hand advantage that refers to the performance of only one hand. Examples of hand ability are that of Braille reading (Millar 1987a) where both hands are used in alternation for reading Braille prose, or actions of typewriting or piano playing. Equal hand ability for both hands has also been found for manual dexterity tasks of sorting or stacking objects, or thumb and finger precision tasks (Ittyerah 2000, 2009).

Handedness has long been a source of fascination and speculation, yet the origin or cause of hand preference is not clearly evident and opinion is still divided on the question of whether or not it is innately determined. Attempts to study the nature and choice of the preferred hand have examined different factors such as social pressure (Watson 1924; cited in Harris 1980), constitutional biases (Collins 1975) and genetic control (Annett 1972). McManus, Sik, Cole, Mellon, Wong and Kloss (1988) have considered whether hand preference is a phenomenon 'sui generis', that is, a result of spontaneous likelihood of the same hand being used, probably due to reasons of efficiency. There are also speculations as to whether hand preferences may have occurred as a consequence of eye- hand coordination (Woodworth 1898; Annett et al. 1979; Peters 1983; Honda 1984). It follows from the aforementioned opinions, that there is a need to discuss the theoretical conceptions of handedness and related assumptions of visual guidance in hand actions. Although the relevance of vision is not explicitly expressed in most theories, none of the theories deny the need for vision in hand actions. The question of interest in this manuscript is to clarify the role of vision in determining hand preference.

Considerable effort has been expended to know about handedness in past populations by examining cerebral asymmetries that can be detected in dry skulls so as to be associated with past prevalence and patterns of human handedness. For example Holloway and de la Coste-Lareymondie (1982) have summarized data from endocasts of hominid skulls to indicate a preponderance of left occipital right frontal width, as well as length asymmetries in humans as early as 2–5 million years ago. In general some of the available evidence suggests that humans may differ from great apes in the prevalence of occipital width asymmetry, which is associated with hand preference suggesting a continuing association between vision and hand actions. However, clinical data from scan studies have shown that occipital width and frontal length asymmetries are evident only in right versus non-right handed women and there are suggestions that the reproducibility of observations of occipital asymmetry is quite low (Chu, Tranel & Damasio 1994).

Therefore consistent relationships between vision and hand preference cannot be determined by examining human skulls or internal organs.

The notion of social pressure claims that environmental circumstances enforce an individual to use a certain hand more often than the other. For example, hand preferences in children differ among cultures and certain actions such as eating or writing are often imposed on the right hand. Komai and Fukuoka (1934) found that less than one percent of Japanese children used their left hand for writing, and Chinese children rarely used their left hand for eating or writing (Teng, Lee, Yang & Chang 1976). Studies of trends in adult populations in Australia (Brackenbridge 1981), Britain (Annett 1973, 1978; Ashton 1982) and in the US (Porac, Coren & Duncan 1980) have documented lower incidences of left handedness with increasing age. However, Annett (1985) observed an underreporting of left handedness in her surveys, thereby suggesting that the documentation of hand preference amongst adults may not be always reliable.

Annett's (1972) right shift theory indicates that the majority of the human population is right sided. In her extensive surveys in Britain, she found that about 66% of the population is right handed, 30% is mixed handed and only 4% is left handed. According to Annett, the handedness of an individual can be determined by the criteria of numbers of hand actions that may be included in a specific category. Annett considered any individual who performed a single action with his/her left hand in a list of other wise right hand performed items to be mixed handed. Therefore Annett (1985) adopted strict criteria of total and complete performance with a single hand to be an indication of hand preference. Not only did Annett find that most people were right handed, but also that women were more right side skilled than men. She attributed this right side preference to be an indication of speech dominance in the left hemisphere in humans and strengthened her argument by showing that the nonhuman species do not have a right or left side preference since they do not possess speech as in humans. Annett claims that the reason for the right shift effect in humans is because humans possess a gene for right hand preference, transmitted by the parents to offspring over generations.

McManus's (1985) theory of handedness is also a genetic theory in which the assumption is that the probability of being left or right handed is equal. According to McManus, every individual has a dextral (D) gene and a chance (C) gene, and the probability of being right or left handed is equal. However, the reason why there are more right handers in any population, is because left handers manifest as right handers because of social desirability. The main difference between the genetic theories of Annett (1985) and McManus (1985) is this: Annett claims that handedness is determined by the skill of the hand with preference being a secondary phenomenon defined by an arbitrary threshold or criterion in a distribution of skill composed of three separate normal distributions, weak right handers centred at zero (showing maximum mixed handers) and medium and strong right handers shifted to the right; McManus believes that hand preferences are a sum of two normal distributions, one to the right of zero and the other placed symmetrically to the left of zero, referred to as the symmetric bimodal model of skill differences (McManus 1985). Both Annett and McManus attribute consistent hand preferences to be related to visual guidance of the hand. Annett claimed that the increasing skill of the right preferred hand is a consequence of lateralized visuomotor proficiency (Annett 1985) and Batheja and McManus (1985) hypothesized that hand writing in sighted schooled children is likely to contribute to increasing lateralization with age.

The importance of the role of vision in task performance was evident as early as Woodworth (1898) when he examined the relation between speed and accuracy of aiming movements. He found that although both hands were equally accurate at reaching the target at slower speeds, the right hand was more accurate at faster speeds both under visual control and eyes closed conditions. In a peg moving task, Annett, Annett, Hudson and Turner (1979) found that though the nonpreferred hand would equal the performance of the preferred hand for placing the pegs in the larger holes, the preferred hand was more accurate at reaching smaller target holes even in visually informed conditions. Peters (1976, 1980) found in finger tapping tasks that the preferred finger tapped faster than the finger of the nonpreferred hand. In conclusion, these authors suggested that the superiority of the preferred side is a consequence of lateralized visuo-motor efficiency. Subsequently Honda (1982, 1984) indicated that the right hand movements toward small targets on the right side were more dependent on eye movements to the right side than the left side and this enhanced the performance of the right hand over that of the left hand. However Annett (1967) observed that the hand preference distributions of human and non-human species differ in that humans alone exhibit a right shift factor, suggesting that a species specific rather than a sensory specific factor is involved.

Other theorists have related hand preferences to head orientation during infancy. Studies dealing with the development of handedness have indicated that one of the earliest findings, referred to as the tonic neck reflex (Gessel & Ames 1947), was found to be partially predictive of later handedness. All the right turning infants became right handed or mixed handed, and about a half of the left turning infants became left handed. Michael (1981) argued that head orientation in infancy is a cause of hand preference, as the infant is more likely to see the hand that is extended on the same side and thereby develop better visuo-motor coordination. An analysis of reaching of infants at 17 to 40 weeks of age distinguished several kinds of movements such as touching by the left hand and grasping by the right hand (Bresson, Maury, Pieraut- Le Bonniec & de Schonen 1977). Ramsey (1980) observed that infants hold the base of a toy with the non preferred hand, while the preferred hand attempted to manipulate it. It was evident that one hand was the more active (Ramsey 1979). Furthermore, there are suggestions that handedness arises because infants have a tendency to adopt an asymmetric posture in their intra-uterine environment, with the hand near the eye so that their right hand is more often in the visual field, enabling easier learning of eye- hand coordination (Peters 1983). If this theory is correct, one should finder a smaller percentage of blind children than sighted children would be right handed. Ittyerah (1993, 2000, 2009) found that the absolute laterality (laterality index) of blind and sighted children is not different either for tasks that were performed mostly by one hand such as writing or pointing or for tasks that required the assistance of the subordinate hand such as cutting with a pair of scissors or winding a string of wool on a needle. The laterality index (LI, McManus et al. 1988) was derived from the performance scores of each child and scores for direction and degree of lateralization were calculated for each child. The scores ranged from −1 through 0 to +1, and those scoring greater than 0 were taken as right handers. The degree of lateralization was considered to be the absolute value of LI (range: 0 to 1). The scores revealed that the distribution of hand preferences for blind and sighted children were not very different (Ittyerah 1993), suggesting that vision is not a necessary condition for determining hand preferences. Most of the blind children like the sighted had a right hand preference and their left hand preferences revealed the typical J shaped curve that Annett (1972) observed in her studies, indicating a gradual decrease in the numbers of tasks performed by the left hand. Therefore lack of sight does not affect the choice of the preferred hand and theories of eye- hand coordination (Gessel & Ames 1947; Michael 1981; Peters 1983; and Honda 1984) are insufficient explanations for the origin of hand preference.

Hand preferences may vary in direction and degree. Direction refers to the choice of the preferred hand and degree refers to the extent to which the preferred hand is used for various actions. Studies of hand crossing the midline of the body have indicated developmental trends in the degree of handedness. Carlier, Doyen and Lamard (2006) observed that when children aged 3 to 10 years were required to reach for cards in the Bishop card sorting task, the older children crossed the midline more frequently when reaching for cards than the younger children. The Bishop card sorting task consists of seven cards that are arranged in a semicircle from the left hemispace across the body midline to the right hemispace. Card 1 is placed at the extreme left of the hemispace followed by cards 2 and 3 in the semicircle and card 4 is placed at the centre of the semicircle and aligned to the midline of the body. Cards 5, 6 and 7 cover the right hemispace so that card 7 is placed at the extreme end of the right hemispace. Carlier et al. found that the effect of the spatial position of the cards was significant. The contralateral hand (hand opposite to a hemispace, for example the right hand in the left hemispace) was used less often to reach cards at the most extreme positions whereas the ipsilateral hand (hand in the same hemispace, for example, the left hand in the left hemispace) was used more often for cards in the extreme positions. The right hand would reach more often for card 7 in the ipsilateral hemispace than for card 1 in the contalateral hemispace. Children that crossed the midline were more lateralized indicating that midline crossing is an act of lateralization. The younger children aged 3–4 years crossed the midline less often than children over 7 years. The spatial position of the cards affected midline crossing. The numbers of hand crossings were higher near the body midline than in the midway or the fartherest positions. The number of midline crossings increased with age and there were more crossings near the midline for every age. There were no differences between left and right handers or gender for midline crossings.

It is quite established that the direction of handedness appears to be fixed by the age of three and the degree of handedness increases over a period of three to seven years and perhaps more slowly from 7 to 9 years in sighted children (McManus et al. 1988). The direction and degree of handedness is not different for congenitally blind children either (Ittyerah 1993, 2000, 2009), indicating that vision does not determine hand preference. Both congenitally blind and sighted children displayed a similar bimodal distribution of hand preference, with most of the children being right handed, and lateralization increased with age in childhood (McManus et al. 1988; Ittyerah 1993, 2000). Therefore there is no evidence that the degree of lateralization in either the blind or sighted children is necessarily a function of learned experience (such as hand writing) with an asymmetric world. Thus the notion that eye- hand coordination is of primary importance in developing hand preference (Peters 1983; Honda 1984) is not evident. If that were the case then the onset of any visual impairment should affect the direction or degree of hand preference. Furthermore, the absence of performance differences between the hands for the blind as well as the sighted children in brick matching, sorting objects, stacking cubes, finger dexterity using the thumb and forefinger and the Minnesota rate of manipulation test (Ittyerah 1993, 2000, 2009) indicated that hand preferences for these tasks may not be evident in childhood even when specific task characteristics (i.e. spatial characteristics) are usually more accurately detected with the left hand both in normal right handers (Witleson 1974) and in left handers (Herrmann & Van Dyke 1978; Porac & Coren 1981). Since the lateralization of the blind children is not fundamentally different from that of sighted children, hand preferences are not confined to sensory experiences and vision neither determines the direction nor the degree of hand preference.

There are suggestions (Lea 1984) that the pre-existing asymmetry of the human brain may have perhaps evolved under the elective pressure of tool use in which the stronger left hand holds the work while the right hand skillfully wields the tool. This may explain why in some studies, fine motor skills that require precision such as writing or sewing are context oriented and better performed by the right hand (Healey et al. 1986; Steinhuis & Bryden 1989; Steinhuis & Bryden 1990; Ittyerah 1996 and Streri 2002), whereas less lateralized tasks such as lifting objects (Steinhuis & Bryden 1989 and Steinhuis & Bryden 1990) showing actions of strength are better performed by the left hand (Healey et al. 1986; Ittyerah 1996 and Peters 1990c). Recently Gurd, Schulz, Cherkas, and Ebers (2006) correlated the hand used for writing with other actions of skill in monozygotic twins with discordant handedness. Gurd et al. observed that although the right handed sisters were more strongly lateralized than their left handed sister, there was no evidence to indicate that twins who wrote with their left hand showed poorer performance than their right handed twin sister. However, certain tasks such as the peg moving did not show evidence of differences between the writing and non-writing hands in the left handed group of mono zygotic twins. All this evidence in sum indicates that general laterality does not affect ability (Ittyerah 1993, 2000, 2009).

Studies revealing the role of the cerebral hemispheres in the control of hand actions have indicated that each hand may be relatively more specialized for particular actions and depend on different processes in task performance. The evidence has generated interest in cerebral lateralization and indicated the need to study the role of both hands during performance. Findings suggested the existence of relationships between the hand and the brain that are controlled by the contralateral hemisphere on each side of the body, showing the importance of studying both hands during an action rather than only hand preference that may be confined to a single hand. Of recent, there is revival of interest in the findings of MacNeilage et al. (1987) who reported evidence of handedness in animals showing distinctions of support and manipulation during hand use. MacNeilage et al. observed that the hands of prosimians showed some specialization such as grasping for support with the right hand and striking prey with the left hand. These findings have been interpreted as signs of laterality in chimpanzees as they use one hand predominantly in bimanual actions such as cracking nuts (Uomini 2009). Uomini found that hand tasks varied in their gradient of task complexity. Simple tasks such as fixing a flint puzzle can be performed equally well with either hand because they are spontaneous unimanual tasks that do not show a hand preference. Other unimanual tasks requiring precision grips such as picking objects express stronger hand laterality and hand laterality is at the extreme of the gradient for complementary bimanual tasks such as nut cracking or retrieving food from a hose/tube like container while the hose is held by the nonpreferred hand and the preferred hand is used to retrieve food from within the hose. Uomini concluded that some tasks that involve coordinated bimanual actions do not elicit the expression of hand laterality (e.g. the flint puzzle task) whereas others such as nut cracking or food retrieving involve complementary bimanual actions and are expressions of laterality. The general consensus is that although humans have three times the brain capacity of apes, greater brain organization complexity and are more lateralized they are prone as apes to a gradient of manual motor complexity. That is, while performing simple tasks such as lifting objects, humans elicit a low degree of hand preference, whereas hand laterality is more significant while performing complex tasks (Mosquers, Geribas, Bargallo, Llorento & Riba 2012; Uomini 2009). Therefore the use of both hands is evident in task performance.

It is of interest in this manuscript to review and relate in the forth coming chapters instances of equal hand ability for motor performances that do not have process related to verbal components. Acts of gestures or pointing that are associated with verbal components have been performed better by the right than the left hand (Jacquet et al. 2011; Esseily et al. 2012) indicating a left hemisphere control for speech in these actions, whereas acts of grasping or reaching have been observed to be performed equally well with either hand indicating that the hands do not differ when a verbal component is not incorporated in performance. These explanations (Umoni 2009; Jacquet et al. 2012) serve as preamble to our findings (Ittyerah et al. 2007) of the ability of the left and right hands to point at targets in the total absence of vision, indicating that though the hands may adopt different orientations during performance, they do not differ in ability.

CHAPTER 3

Hand and brain

The relation of hand and brain has been of increasing interest in neuropsychology since somatosensory laterality experiments revealed behaviours that reflect functional control of the cerebral hemispheres with the hands. Traditional experiments using hands in somatosensory research have presented stimuli to either hand to test for detection of stimulus characteristics. It has long been known that speech functions are primarily localized in the left hemisphere of normal right handed individuals and certain aspects of spatial and musical abilities are localized in the right hemisphere of right handers. The general consensus is that damage to the left hemisphere results in language disabilities in most people. Anatomically, the ascending sensory fibres from the skin and joints cross the midline and project to the postcentral gyrus of the opposite hemisphere. There is also sensory representation in the precentral gyrus, the traditional motor cortex (Rasmussen & Penfield 1947). Besides there is a second sensory area, along the superior border of the Sylvian fissure in the parietal cortex, in which both ipsilateral and contralateral sides of the body are represented in each hemisphere (Penfield & Rasmussen 1950). There is also evidence for bilateral representation of the lower part of the face area in the postcentral region (Corkin 1978). Thus the somatosensory system does not contain only contralateral sensory fibres because some information is projected via ipsilateral pathways (Albe-Fessard 1967; Semmes 1968).

3.1 Vertebrate lateralization

Studies of vertebrate lateralization with birds and mammals have indicated left and right hemifield advantages showing that brain lateralization is fairly similar across phylogeny. Comparative studies clarify conceptions of structure and function of brain behavior relationships in humans by drawing similarities in fundamental sensory and motor processes in different species. Lateralization may be generated by mechanisms common to more than one sense modality. For example if a bird turns its right eye to grain on the floor, it is also likely to be accompanied by its right ear. The role of attention is important since attending to a visual target may also be accompanied by auditory attention in the same hemifield. Just as the left- right mechanisms for visual processing are fairly established, there is also left -right specialization of perceptual processing for hearing and touch (Andrews & Watkins 2002) that involve the left side of the fore brain in the control of responses to a target and the mechanisms on the right side of the forebrain for spatial relations.

Dichotic listening tests in humans have shown that the ear to which attention is directed is processed faster (Hugdahl 1995) with right ear advantages in most instances, particularly for verbal stimuli. Attending to the perceived position of sounds has also been found in animals like the Japanese macaque revealing a right ear advantage for discriminating variants of species specific vocalization (Petersen et al. 1978). Ehret (1987) found that when either the right or left external meatus of a lactating mouse was blocked, she would approach the calling pups only when the right ear was open. Similarly McKenzie (1991, cited in Andrew & Watkins 2002) observed that chicks are unable to differentiate a cluck from other sounds when the meatus contralateral to the controlling hemisphere is plugged. Thus the right ear use results in left hemisphere involvement in the analysis of conspecific vocalizations (Andrews & Watkins 2002).

Tactile lateralization has been little investigated in animals. Though relatively less investigated as compared to vision or audition, there is good evidence to show similar effects of lateralization in touch as was found in vision or audition. LaMendola and Bever (1997) studied rats in a maze task with a baited arm. The left or the right whiskers of the rats were anathetized before the rats were allowed to enter the maze. There were fewer errors in finding the baited arm when the left whiskers were anasthetized, indicating the involvement of the right whiskers in locating the bait. Left hemisphere involvement in the control of the right whiskers was confirmed by unilateral spreading depression of the left or right cortex when both whiskers were used. Left hemisphere depression in the rats led to more errors in the maze task. The authors concluded that there is consistent lateralization for the use of whisker information in rats for the discovery of food.

3.2 Lateralization of object recognition

Studies in object permanence have shown infants at about 18 months are able to represent objects in their absence (Piaget 1953) though subsequent studies have indicated that representation of objects may be possible at the early age of 2 months (Baillargeon 1994) even before infants are able to develop action schemes to reach the object. This change in perspective has allowed scientists to conduct comparative studies in precocial birds like chicks to know the extent to which chicks are able to recognize objects they have seen that are partially occluded. Perception of object unity in partial occlusion displays have been demonstrated in two month old human infants (Kellman & Spelke 1983; Johnson & Aslin 1995). Among mammals recognition of occluded objects has been found in mice (Kanizsa et al. 1993) and monkeys (Osada & Schiller 1994). Vallortigara and Regolin (2002) found that chicks after training were able to choose a complete triangle from a triangle with a section removed. Chicks continued to choose the complete triangle even after it was occluded by another triangle that had a portion removed from it, though both triangles contained the same amount of red and black areas. These effects were also demonstrated in adult hens (Forkman 1998) when hens pecked significantly more often at the stimulus that was occluded, showing that they are able to recognize the occluded object as a complete one. Chicks and hens possess visual abilities similar to those of human infants for visual perception of partly occluded objects. Lateralization of brain functions has been demonstrated in domestic chicks (Rogers 1982, 1989, etc) where the left eye system/right hemisphere is inclined to a global analysis of the visual scene and the right eye system/left hemisphere seems to be inclined to feature analysis (Vallortigara & Regoloin 2002).

The ability to represent completely occluded objects in their absence has been found among chimpanzees, gorillas and dogs (Etienne 1974; Dumas 1992). In a detour task when chicks had to go around a barrier to reach a goal, Vallortigara and Regolin (2002) found that the right or left direction of detour around the barrier reflected preferential fixation by the contralateral eye during the detour, that is for a detour on the right the preferential left eye was used and for detour on the left, the right eye was used. Chicks using the right eye took less time to detour the barrier than chicks using the left eye. Binocular chicks showed a preference to detour the barrier on the left side and were similar to chicks that used the right eye to locate the target earlier than those using the left eye. There are also indications that chicks are able to differentiate 'where' the target is from 'what' is the target, showing an integration of information that involves neural structures in the left hemisphere (Cozzutti & Vallortigara 1999, 2001).

Hand use and signing behavior in primates usually obtained from small samples of chimpanzees or gorillas have not yielded consistent findings for a side bias. Hopkins and Leavens's (1998) study on manual gestures in chimpanzees is one of a few that has reported population level right hand bias in manual gestures for food begs contrasted with whole hand pointing. The authors also noted that right hand use for manual gestures was more frequent when it was accompanied by vocalization than when it was not. Lateral biases in face recognition and discrimination in monkeys were found by Hamilton and Vermeire (1983) who reported right hemisphere superiority for discriminating conspecific individuals and their facial expressions in split brain monkeys. Vermeire and Hamilton (1988) reported similarities between humans and monkeys when they found the expected right hemisphere dominance in split brain monkeys for faces in normal positions though not in their inverted positions. Examples of global and local processing of visual stimuli have shown that chimpanzees and baboons showed a right visual field advantage for local processing and a left visual field advantage for global processing (Hopkins 1997; Deruelle & Fagot 1997; Fagot & Dreulle 1997) as in human subjects. Evidence indicates that upright faces were encoded wholistically, whereas inverted faces were encoded in a piecemeal manner (Yin 1969, Carey & Diamond 1997). One may infer from these finding that the right hemisphere dominance for faces in humans and monkeys could be an outcome of encoding faces wholistically.

Behaviour asymmetries have also been found in the processing of conspecific vocalizations. Petersen et al. (1978) found that Japanese Macaques were able to discriminate conspecific vocalizations better with the right ear (left hemisphere processing) than their left ear (right hemisphere) indicating a communicative valence in the calls. Hauser and Anderson (1994) found that rhesus monkeys turned their right ear to conspecific calls and the left year to an alarm call. The perceptual asymmetries for vocal perception are supported by both neuroantomical and experimental lesion studies. In studies with humans showing left hemisphere biases for language processing, the Slyvian fissure is longer in the left hemisphere than in the right hemisphere (Geshwind & Levitski, 1968). It is assumed that the length of the fissure corresponds to the auditory cortex. Similar asymmetries have been found for apes (Le May & Geshwind 1975), old world monkeys and new world monkeys (Falk et al. 1986; Heilbroner & Hollowway 1998) for left Slyvian fissure length. Thus the perceptual asymmetries observed in the laboratory (Petersen et al.) and external conditions (Hauser and Anderson) are associated with these anatomical asymmetries. Furthermore homologies in the language areas of apes and humans have been observed. Just as humans have a larger planum temporal in the left hemisphere, a left-right asymmetry in the planum temporal has been found for chimpanzees and other great apes (Gannon et al. 1988a, 1988b). Furthermore there is behavioural and acoustic evidence in old world monkeys that the spectral properties of their vocalizations are a part of their articulatory gestures such as movement of the lips, tongue and jaw during vocalization (Hauser, Evans & Marler 1993; Hauser & Schon Yharra 1994). Thus cerebral asymmetries observed in humans have been found in animals as well, indicating the relevance of comparative evidence in brain-behaviour relationships.

3.3 Somatosensory asymmetry

The conventional understanding of somatosensory asymmetry was that the crossed representation of the somatosensory system would lead one to expect greater right side advantages for verbal tasks because of the greater involvement of the left hemisphere, and left side advantages for nonverbal tasks due to the contralateral control of the left side of the body by the right hemisphere. Many of the initial studies of somatosensory asymmetry involved measures of pressure sensitivity with von Frey hairs. For example, Weinstein and Sersen (1961) reported that right handers were more likely to show lower thresholds on the left palm, forearm and sole than on the right side. Left handers with a familial history of left handedness had lower thresholds on the right side of their bodies, whereas left handers without a familial history of left handedness behaved like right handers. The distinction between familial and non familial left handers showed that these effects were due to some central asymmetry. The general trend toward greater pressure sensitivity on the left side recurs in other studies (Weinstein 1962, 1963) while some others fail to show these differences (Fennell, Satz & Wise 1967; Weinstein 1968). Therefore the differences between the left and right sides in pressure sensitivity seem to be small and inconsistent.

Measurements of two point thresholds are not very different. Weinstein (1963) found a lower two point threshold on the right breast, but later measurements (Weinstein 1968) indicate that this does not hold across different body parts. Even if there are indications of a better two point limens on the left side of the tongue (McCall & Cunningham 1971; McNutt 1975), these findings were not statistically significant. Corkin (1978) in a review concluded that there is no evidence for any lateralization effect in basic somatosensory measures. Since these are simple low level tasks not requiring any complex cortical processing, they may show lack of laterality effects. To find somatosensory effects related to cerebral asymmetries, there must be more complex verbal and nonverbal tasks.

The traditional experiments using hands in somatosensory research have presented stimuli to either hand to test for the detection or recognition of stimulus characteristics. This is analogous to the unilateral visual presentation technique, and successive experiments have presented stimuli to both hands for simultaneous inspection. Young and Ellis (1979) found left hand superiority for judging the number of random dots in an array. Subjects felt an array of raised dots with their middle finger, working rapidly through a block of ten trials. This left side superiority for dot patterns was similar to the left hand superiority reported for reading Braille letters (Hermelin & O'Connor 1971; Rudel, Denckla & Hirsch 1977; Smith, Chu & Edmonston 1977). Left hand superiority has also been reported for the tactual determination of direction. The subject was required to feel one or two metal rods embedded in wood with the fingertips of one hand and then select from a visual display the rod(s) that have the same orientation. In general right handed subjects were more accurate on the task with their left hands (Benton, Levin & Varney 1973; Varney & Benton 1975). Varney and Benton (1975) also found that familial left handers tend to show a right hand superiority. Benton, Varney and Hamsher (1978) showed that the same results could be obtained with a tactual matching task. Most importantly, the task seemed to be sensitive to lesions of the right parietal lobe (Carmon & Benton 1969; Fontenot & Benton 1971). Right hemisphere damage produced bilateral impairment on the task, whereas left hemisphere damage resulted in an impairment of the left hand. Evidence for left handed superiority in tactual form recognition was obtained by Dodds (1978) and Hatta (1978) indicating more right hemisphere involvement in form perception. These unimanual laterality effects were left handed (right hemisphere) effects, and there was little evidence for right hand superiority with verbally related tactual tasks.

Witelson (1974) was perhaps the first to attempt a dichhaptic analogue to the dichotic listening task. Dichotic listening involves the simultaneous presentation of two different auditory stimuli such as two different words to the two ears and the subject is required to report what s/he heard. In dichhaptic presentation, Witelson had children palpate two nonsense forms one with each hand simultaneously. Subjects then selected the two forms that they were required to palpate from six visually presented stimuli. Boys from grades 1 to 8 showed a left hand advantage for the forms and a right hand advantage for pairs of letters they had felt similarly. In a later study Witelson (1976) found that girls did not show the left hand advantage for haptic form until the age of 12. Oscar-Berman et al. (1978) presented two tactual stimuli simultaneously to the palms of the hands, using a passive presentation procedure in which the figure was traced on the hand. Subjects were required to identify the two items that had been presented in a specified order, with the left hand being given first on some trials and the right hand first on other trials. They found right hand superiority for letter identification, no difference for digit identification and a left hand effect for line orientation.

Nachshon and Carmon (1975) tested subjects both unimanually and bimanually on tasks they described as 'sequential' and 'spatial'. In their sequential task the index, middle, and ring fingers were stimulated in random order by tapered metal rods, the three stimuli arriving at .5 second intervals. Subjects then had to press micro switches next to the rods in the same sequence as they had been stimulated. In the spatial task one finger was stimulated once, one finger twice and the third finger not at all. Again the subject had to indicate the pattern of stimulation by pressing the appropriate micro switches. For bimanual presentation, only the thumb and middle finger of each hand were used. In the sequential task, all four fingers were stimulated in random order, while in the spatial task, one finger was stimulated twice and the other finger of the other hand, not at all, and the two fingers of the other hand were stimulated once each. Separate groups of 20 right handed subjects were tested on the four tasks. With unimanual presentation, no significant asymmetry was observed, although there was a trend toward fewer errors for the right hand in the spatial task, and significantly more subjects were better with the right hand than the left. With bimanual presentation, subjects made more errors with the left hand for the sequential task and more errors with the right hand for the spatial task. As in dichotic listening, the right hand advantage found for the bimanual sequential task would tend to implicate the left hemisphere in sequential analysis. The study indicated dissociation between spatial and sequential tasks observed under conditions of competition. The general pattern that emerged from the somatosensory studies was compatible with those in audition and vision. A left side, right hemispheric advantage was found with spatial tasks that cannot be easily verbally mediated, such as discriminating line orientation, whereas a right handed left hemispheric superiority was found for verbal mediation (letters) or fine temporal discriminations as in the sequential tasks. The line orientation task was critically dependent on the integrity of the right parietal lobe (Fontenot & Benton 1971).

3.4 Left and right: Brain and hand

Handedness is an asymmetry that has long been known and studied. The incidence of left handedness in the population is generally considered to be between four to ten percent although the specific figures vary considerably, as a function of the source of the sample and the procedure for measuring handedness. The majority of studies of the incidence of left handedness have used rather crude measures of handedness, such as asking subjects for their writing or their preferred hand. It is well known that speech functions are primarily localized in the left hemisphere of normal right handed individuals (Corballis 2003) and certain aspects of spatial and musical abilities are localized in the right hemisphere of right handers (Hecan & Albert 1978; Milner 1974). Satz (1980) in his review on aphasia in left handers concluded that speech is bilaterally represented in many left handers. His survey indicated that aphasic disturbances were found following unilateral brain damage more often in left handers than right handers and he concluded that about 40% of left handers have bilateral speech representation.

The development of sodium amytal techniques for the assessment of speech lateralization (Wada & Rasmussen 1960) had made it possible to obtain data on the relation between handedness and speech lateralization from large samples of subjects. Studies employing this technique revealed that the majority of left handers (70%) like right handers (90%), had speech in the left hemisphere (Milner, Branch & Rasmussen 1964; Rossi & Rosadini 1967), although the data were discordant with those presented by Satz (1980). Yet another approach to estimating the distribution of speech lateralization had been to investigate the incidence of dysphasia following unilateral electro convulsive therapy (ECT) in patients undergoing treatment for depression, but without signs of organic disorders. Among 30 left handers, left hemispheric speech was reported in 70%, right hemispheric speech in 23%, and bilateral speech in 7%. There seemed to be a general agreement in the figures arising from different techniques of assessing lateralization (Rasmussen & Milner 1977; Warrington & Pratt 1973). One possibility, that left handers with a familial history of left handedness show reversed hemispheric specialization, was indicated with related techniques as dichotic listening and visual hemifield procedures (Zurif & Bryden 1969). In a clinical population, Hecaen and Sanguet (1971) found evidence to suggest that it was familial sinistrals who show reversed functional specialization. The relationship between handedness and brainedness has also been found in cases of dysphasia following brain lesions. Zangwill (1967) found the highest risk of dysphasia in right handers with left sided lesions (60%), but the risk to left handers with left sided lesions was not significantly lower (55%). The main difference between the right and left handers was in the risk of dysphasia following right sided brain lesions: about 2% in the right handers and 29% in the left handers. This comparison showed a difference between the two handedness groups for brain laterality, though there was little clarity about the nature of the association between handedness and brainedness.

Observations conveyed ipsilateral brainedness for speech in nonright handers. Temporary dysphasia was more often found to occur in left handers (Hecaen & Piercy 1956). In right handers the dysphasia was usually associated with left side lesions, whereas in left handers, the disturbance of language occurred whether the left or the right side had the lesion. These findings suggested that left handers with either hemisphere damaged have a higher risk of speech disturbance than the right handers and therefore possibly both hemispheres may be involved in the language function of left handers. Brain images of left and right handers (LeMay 1977) indicated that the brains of the left handers differ from that of the right handers in that the brains of the left handers are not the opposite of right handers, but that they are about evenly divided for biases of planum temporale and carotid artery asymmetries to each hemisphere.

The general consensus is that damage to the left hemisphere result in language disabilities in most people. Among the symptoms associated more often with the left than right side lesions in right handers are disorders of speech production and comprehension, reading, writing and calculation (Hecan & Sanguet 1971).

Intelligence test scores are also negatively affected by left than right side lesions (Meyer & Yates 1955; Smith 1966). In addition to these tasks left side lesions are also associated with disorders of right-left orientation, recognition of colours and the ability to perform some learned actions (apraxia) that depend on the left hemisphere such as brushing teeth or stirring coffee.

Lesions of the right hemisphere are not associated with dramatic loss of function such as an inability for speech production after lesions in the left hemisphere. One of the symptoms associated with right cerebral lesion is neglect of the left half of space so that the patient may ignore the left half of a picture or draw only the right side of a house. Unilateral neglect extends to all sensory modalities and also affects the imagined representation of space. Bisiach, Capitini, Luzzatti and Perani (1981) found that when patients with unilateral neglect were required to describe images of the Cathedral of Milan they had seen with their right eye controlled by the left hemisphere, they displayed a residual capacity to recall what may have been perceived by the left eye controlled by the damaged right hemisphere. Their descriptions involved the right side view of the Cathedral seen by the left eye. Another symptom found among patients with right hemisphere lesions was dressing apraxia or an inability to wear clothes, as well as a difficulty in localizing objects in space (Brain 1941). Patients with right side lesions also showed difficulty in locating points in space or in discriminating the shape of blocks by touch (De Renzi 1978, 1982), indicating in general that the right hemisphere plays a special role in the apprehension of spatial relations, independently of sensory modality (Benton 1985). Patients with right posterior lesions have difficulty in recognizing pictures of letters, objects and faces when the representation is degraded or confused by the overlap of drawings (De Renzi & Spinnler 1966; Warrington 1982). Patients with right temporale lesions were found to be impaired on tests of spatial learning and memory (Milner 1974) and on tests of musical perception (Milner 1962). Lesions to the right frontal lobe were associated with difficulties in the planning of movement sequences and also of keeping track of the order in which pictures were seen (Milner 1982).

Although in general most studies indicate that the right hemisphere is specialized for visuospatial abilities and the left hemisphere is dominant for language, there has been little investigation of the relation of handedness to the right hemisphere. The role of the right or the minor hemisphere in higher cognitive processes has shown several nonverbal functions to be mainly controlled by the right cerebral cortex. Among these is the ability to visualize the total configuration of a stimulus from partial information. Right hemisphere lesions have been correlated with defects on tests such as the Mooney faces (Lansdell 1968), street completion tests (De Renzi & Spinnler 1966), and Gollin figures (Warrington & James 1967), which require identification of a figure most of whose contour is missing. Hecaen, DeAgostini and Monzon-Montes (1981) reporting on the incidence of spatial disorders as a function of unilateral brain damage found that spatial disorders are more likely to follow upon right hemispheric damage, but that both sex and familial history of sinistrality affect this relationship.

3.5 Dissociation of thought and action

Research studies that gained the attention and fascination of hand and brain relationships were initially conducted by Dr. P.J. Vogel and his staff at the White Memorial Medical Centre in Los Angeles with epileptic patients following cerebral commissurotomy (Bogen & Vogel 1962; Bogen, Fisher & Vogel 1965). Patients suffering from epileptic seizures, having endured intense pain, had undergone cerebral commisurotomy to relieve intractable epilepsy. The operations involved complete section of the corpus callosum, the anterior commisure and possibly the hippocampal commisure. A summary of conclusions to human cerebral specialization from studies of split brain patients was given by Sperry (1974). The left hemisphere was considered to be the main language centre, while the right hemisphere was credited with only simple language comprehension. Smell was lateralized to each nostril but the olfactory nerves were uncrossed, such that the patients can talk about smell presented to left but not the right nostril. Each hemisphere was aware of the contralateral hand and visual field. Auditory messages were received from both ears, but those of the contralateral ear took precedence when there was competition.

The right hemisphere was assigned spatial abilities and non-verbal ideation. Simple figures like rectangles were drawn better with the right hand in right handed split brain patients, but the spatial relationships of more complex figures like a house were better represented by the left hand (Gazzaniga & Sperry 1967). The left hand (right hemisphere) was able to arrange three dimensional blocks to match a pictured design when the right hand could not (Gazzaniga 1967). More specifically, the right hemisphere in commisurrotomy patients was superior to the left in judging the size of a complete circle from which a given arc had been taken (Nebes 1971). In instances in which the right hemisphere had been retracted for right handed patients, the accuracy of the left hand over that of the right was found for selecting the appropriate alternative in tests of figure matching. When subjects were required to infer the overall shape of a figure from examination of its disconnected pieces and identify an appropriate tactile part from a set of three alternatives the accuracy of the left hand was very high for almost all the right handed patients while that of the right hand was at chance (Nebes 1972). Since haptic information from each hand projects to the contralateral hemisphere, the discrepancy in scores between the left and right hands indicated an asymmetry in the capacity of the two sides of the brain for this part-whole task, with the right or the minor hemisphere being markedly superior. Split brain patients were better able to manipulate patterns made from bent wire and match them with visually presented shapes with the left hand controlled by the right hemisphere (Milner & Taylor 1972) and subsequent research demonstrated that the right hemisphere has a special role in solving geometrical problems (Franco & Sperry 1977).

Experiments with split brain patients (Sperry 1974) revealed the independent role of each hemisphere in the control of specific behaviours. When a split brain patient was presented a verbal stimulus in the right hand, he was able to name it correctly and when presented with a nonverbal stimulus in the left hand, he was able to identify it correctly. This indicated the control of the dominant contralateral hemisphere for a particular behaviour. A curious phenomenon was found when they were presented with chimeric figures. A chimeric figure is one that has one half of one figure joined to the other half of another figure, for example, half an eye on the left and half a flower on the right, so that the whole picture is seen as half eye with half flower. When a split brain patient was presented such a chimera to the center of his visual field, his right eye would claim to have seen only the flower as a whole, although his left hand would be able to identify the eye on the left half of the picture from several other stimuli. Similarly, his left eye would claim to have seen only the eye as a whole figure, while his right hand would be able to pick the flower from a set of other stimuli. This indicated the dominance of each hemisphere for stimuli presented on the contralateral side and the implicit involvement of the ipsilateral hemisphere. The shift to right visual field stimuli was especially clear when the task was to choose from one set of pictures (toes, pie, key) the one whose name rhymed with pictures presented chimerically (rose, eye, bee) (Levy & Trevarthen 1977). Only the left hemisphere of most patients tested could match the pictures presented for the sound of their names, indicating that only the left hemisphere could generate an internal representation of the word sound. Zaidel (1976, 1977, 1978) in a series of studies observed that the right hemisphere was able to perform nearly as well as the left hemisphere in simple language tasks but may be poorer than the left hemisphere in phonetic discrimination of words in a sentence sequence. Zaidel concluded the right hemisphere is able to understand grammatical and semantic relationships in words and phrases, though it has difficulty understanding long sentences, and therefore the right hemisphere has its own pattern of language skills. Gazzaniga and Ledoux (1978) argued that neither the left nor the right cerebral hemisphere is wholly specialized for any one function, and at best the hemispheres function in an integrated manner. Asymmetries have also been observed in normal subjects to stimuli presented to each side of the body. Early examples are that of dichotic listening (Broadbent 1956; Kimura 1961) where most people were observed to have a right ear advantage for lists of words presented simultaneously to both ears.

Other instances of dissociations in visual consciouness have been reported with clinical patients, as in blindsight (Weiskrantz 1986) where consciouness for the shape of the object was denied, though the ability to point to the correct location was intact. Similarly dissociation for touch was found (Paillard, Michel & Stelmach 1983) when a patient was unable to report what object had touched her arm, though she was able to point accurately at the place on her arm that was touched. Early publications about blindsight described patients with lesions of the primary visual cortex who exhibited remarkable visual capacities to orient their gaze or direct their hand toward targets presented within their blind hemifield (Weiskrant, Warrington, Sanders & Marshall 1974; Perenin & Jeannerod 1975). This residual function was attributed to subcortical vision, following the distinction in animal experiments (Schneider 1969). The lack of awareness implied that patients usually felt like they were guessing, and was compatible with the idea that subcortical vision is unconscious. However the two visual systems model was challenged by other findings (Milner & Goodale 1993; Jeannerod & Rossetti 1993). There was another conception of vision as a dissociable function that considered both modes of vision as mediated by corticocortical pathways: the 'where' function would depend on a dorsal stream projecting from the primary visual cortex to the posterior parietal lobule, and the 'what' function on a ventral stream from the primary visual cortex to the inferotemporal cortex (Mishkin, Ungerleider & Macko 1983). This opinion has been further modified and now it is argued that the inferior parietal lobule of primates, rather, provides a set of modules specialized for visually directed action, whereas the inferotemporal cortex is primarily concerned with object recognition. The dorsal pathway would thus be concerned with pragmatic motor representations about 'how' to act toward an object, and the ventral pathway would be involved in building more semantic representations about 'what' the object is as an identifiable entity (Goodale & Milner 1992; Jeannerod & Rossetti 1993; Milner & Goodale 1993; Jeannerod 1994).

The functions of the two cortical pathways have been observed in patients with optic ataxia, following a lesion of a restricted area of the posterior parietal lobule. The patients were found to have difficulties in directing actions to visual objects, although they were not impaired in the recognition of these objects (Perenin & Vighetto 1988; Jakobson, Archibaldi, Carey & Goodale 1991). They exhibited deficits not only in their ability to reach toward the object, but also in adjusting the hand orientation and shape during reaching. These results suggest that the posterior parietal cortex plays a crucial role in the organization of object oriented actions, whether the visual processing required for a given action is concerned with spatial vision (location) or with object vision (size or shape) (Jeannerod & Rossetti 1993). Goodale, Milner, Jacobson and Casey (1991) observed dissociation that complements the above finding in a patient (DF), who developed a visual form agnosia following a bilateral lesion of the occitotemporal cortex. Despite her inability to perceive the size, shape and orientation of visual objects, DF performed quite accurately when instructed to perform movements toward these objects. This observation suggests that during action, DF could still process visual information about the object's intrinsic properties she could not perceive. Optic ataxia and visual agnosia patients make good cases for the dissociations between recognition of objects and actions oriented toward objects.

Although various residual functions have been reported in cortically blind hemifields, the majority of them are related to extrinsic properties of objects, that is mainly location and motion (Weiskrantz 1989; Adams et al. 1990). Perenin and Rossetti (1996) tested patients with V1 lesions to know if they were able to unconsciously process orientation, size or shape of visual stimuli during action. In this manner they expected to generalize their findings to the abilities of blindsight patients. The patients were presented with slots of variable orientations or with rectangular objects of equal surface having variable horizontal lengths. Their performance was assessed in a verbal, motor and a matching task. In the verbal task they were required to produce forced choice verbal guesses about stimulus orientation or size. In the motor task they had to insert a card into a slot with the appropriate orientation or grasp the rectangle between thumb and index finger. In the matching task they were required to match the orientation of the slot into which the card was to be placed by wrist prosupination movements or to match object horizontal size holding the card between the thumb and the forefinger. One patient (PJG), 32 years, with a complete right hemianopia due to a left medial occipital lesion was able to discriminate motion direction in his hemianopic field (Perenin 1991), but was unable to discriminate between simple geometric forms of circles and triangles. When instructed to perform each of the three tasks with his left hand in the normal visual field, he performed as well as healthy subjects for either of the two types of stimuli. When required to perform on the right side, the verbal and the matching responses were at chance. However, there was a significant relationship between the slot and the hand orientations for the reaching responses ($r = .43$, $p < .005$). Similar effects were obtained when PJG grasped the horizontal objects. While he performed randomly in the verbal and matching tasks, his maximal finger grip aperture (measured during the transport phase) and his final grip aperture (measured at the time of contact with the horizontal panel) were both significantly correlated with the actual object size. Another patient NS was able to introduce the card into the slot, to her own surprise, (9 out of 40 trials) without being informed that only four orientations were being used)

(Rossetti, Regnier, Perenin, Rode, Lacquaniti & Boisson 1995b). These data provide another instance of dissociation between two modes of visual processing (knowing what the object is versus how to grasp it). They indicate that the neural pathway for space representation in action (pragmatic representation) is much less dependent on V1 input than is the pathway involved in visual discrimination, identification and visual awareness.

In an earlier study Rossetti, Rode and Boisson (1988) tested a patient JA with a lesion in the thalamic relays of somaesthetic afference. The tactile and proprioceptive deficit was complete on the right side of the body (lesions of left ventrolateral and ventroposteriolateral nuclei). Firstly JA was tested for his ability to locate tactile stimuli applied to his right hand. The patient was blindfolded and motor and verbal performances were compared. Motor responses involved pointing movements using the left index finger. Verbal responses were obtained using a forced choice paradigm where the patient chose between stimulus locations that were shown to him before the test. When he responded by pointing with the left hand to the stimulated right arm (pointing on arm condition), he consistently performed well above chance level. The verbal forced choice paradigm demonstrated that the deficits exhibited by JA could not be simply explained by a conscious- unconscious dissociation, since verbal guesses were at random. Secondly, in order to test whether the somatic sensation was processed only for the motor interaction with the stimulus, or if it could also locate the perceived sensation, the drawing of an arm was placed on the Table 20 cm left of the patient's hidden stimulated right arm. JA was then asked to point on the drawing to the point matching the location of the stimulus applied to his right arm. In this condition he had to know 'where' to point, instead of only knowing 'how' to point, as in the instances of the pointing on arm condition. Comparison of the two conditions showed that the patient could not perform accurately in the pointing on drawing condition.

Rossetti, Rode and Boisson (1998) tested patient JA for his ability to process proprioceptive information. As for touch, JA was blindfolded and his right numb arm was manipulated by the investigator in such a way as to place the right index finger on one of two locations on a table, either at the right position or the left position. Since JA was not able to consciously process any proprioceptive information about the arm, he was evaluated for two responses. In one session JA was asked to point underneath the table to the position corresponding to his right target fingertip position. The index finger could be at the right or the left target position on the table. JA was significantly influenced by the target finger locus in the pointing task. In the second session when JA was required to simultaneously point to the target and label it verbally, his performance dropped to chance level and he could no longer discriminate between the two loci (pointing + verbal)

Simultaneous activation of the pragmatic and semantic representations produced a deleterious effect on the motor response.

The failure of these subjects to perform above chance in the verbal forced choice conditions show that sensory information (visual, tactile, proprioceptive) may be processed, not only implicitly, but also specifically for motor purposes. Therefore these findings can be interpreted as dissociation between a motor system responsible for the stimulus driven pointing and a semantic system responsible for the verbal depiction of the same stimulus. A similar dissociation had been previously reported between a 'what' system responsible for semantic processing and a 'how' system responsible for pragmatic processing (Goodale & Milner 1992). It is now accepted that the posterior parietal cortex is primarily involved in visual processing for action purpose (Sakata & Taira 1994). In contrast, the dissociation observed in JA applies to stimulus location (What vs How) and is applicable for both touch and proprioception. This view was strengthened in the instances when JA pointed on the arm drawing. In this event JA had to produce a pointing movement combined with an elaborated representation of where the stimulus has been applied, and this in turn reduced his performance to chance level. The dissociation observed with JA would therefore not result from the difference in the response provided (pointing vs verbal), but from the difference between the representations underlying the responses (How vs Where).

Neuropsychological evidence shows the existence of two distinct visual systems devoted to action (Pragmatic or 'How' system) and perception (semantic or 'what' system) respectively (Goodale & Milner 1992; Jeannerod 1994; Jeannerod & Rossett 1996). Patients with parietal lesions have been found to be able to describe objects they are unable to grasp (Perinin & Vighetto 1988) and conversely patients with temporal (Goodale, Milner & Jakobson 1991) or occipital (Perenin & Rossetti 1993) lesions have been found unable to describe objects they can grasp. Thus the pragmatic and semantic systems in these patients are controlled by two pathways. Recent evidence shows that when optic ataxic patients were instructed to reach a target after a delay (Milner, Dijkerman, McIntosh, Rossetti & Pisella 2003) performance improved as compared to immediate performance because of the benefits of pointing on the basis of stored location information. When reaching rapidly toward targets, these patients cannot like healthy subjects use the dorsal stream visual motor system and therefore make large errors in pointing at targets. This is because the effects of brain damage do not provide ready access to the visuo motor system controlled by the dorsal stream for accurate immediate reaching. The authors proposed that for both immediate and delayed reaching, the ventral stream provides the visual information to programmed reaching movements in optic ataxic patients and not the dorsal stream as in healthy subjects. In normal subjects the faster dorsal stream preempts immediate reaching (Rossetti 1998; Rossetti & Pisella 2002) and therefore the ventral stream has little influence on the visuomotor control of immediate reaching.

There is also some evidence that dorsal stream functions that are proposed to compute vision for space perception and the control of actions are shaped by nonspatial information early in development. Perceiving space and guiding movements are not only based on vision, but also on other sensory modalities such as proprioception and kinaesthesia. Experiences of early blindness provide exceptional examples to study the plasticity of the dorsal stream functions. Fiehler and Rosler (2010) investigated action control and space perception in congenitally blind and sighted adults while performing active and passive hand movements without visual feedback using psychophysical and fMri methods. The functional imaging data showed largely overlapping activation patterns for kinesthetically guided hand movements in congenitally blind and sighted subjects covering regions of the dorsal stream. In contrast to the sighted the congenitally blind additionally activated the extrastriate and auditory cortex. Furthermore, the performance of the congenitally blind on the psychophysical task was correlated to their prior experience at mobility training and orientation. These findings suggest a multisensory network of movement control that develops on the basis of sensory motor feedback rather than being under the exclusive control of vision. Thus visual deprivation seems to lead to both cross-modal and compensatory intramodal plasticity. These findings also imply that the dorsal stream functions are shaped by nonvisual spatial information early in development.

The dissociation between the ventral and dorsal stream pathways has also been demonstrated in normal sighted and blindfolded sighted subjects with intact brains when responses to reach or point at the stimuli are guided by immediate or delayed instructions.

3.6 Effect of delay in thought and action

An object and its orientation are located in one or more frames of reference. There are two major frames of reference, the egocentric frame of reference and the allocentric frame of reference. In vision both egocentric and allocentric frames of reference coincide as we locate objects or move in space. In blind conditions subjects rely on an egocentric frame of reference since the allocentric reference such as a door or the wall is not visible. A reference frame gives the physical location relative to an objects position and orientation. In an egocentric reference frame objects are represented with respect to a particular perspective of the perceiver, whereas an allocentric reference frame locates objects within a frame work external to the perceiver such as a wall and is independent of his position. Different body parts have been identified to serve as egocentric frames of reference. For example in haptic perception the hand (Kappers 2004; Volcic 2008) and the body midline (Millar & Al-Attar 2004) have been a good source of egocentric reference. In memory guided movement paradigms (Rossetti 1998) with instructions to point at targets after delays, pointing responses to immediate instruction generate pointing scatters that represent the adoption of an egocentric frame of reference, whereas pointing responses to delayed instruction generate pointing scatters that represent an allocentric frame of reference.

One interpretation of the effect of delayed instruction to point at targets is that the pragmatic representation establishes a direct connection between stimulus and response thereby bypassing the semantic representation of the stimulus. However, action could also rely on the semantic system, once the time required to build a semantic representation of the system has elapsed. Rossetti, Rode and Boisson (1995) asked patients to provide a verbal response about stimulus location (e.g. right versus left) during the course of their pointing movements. It was assumed that a semantic representation of the stimulus was built under this condition. Patients tested for vision, touch and proprioception provided the same results: movements to undetected stimuli became nonadapted when simultaneous verbalization was required. These results show that a pragmatic system can produce movements to stimuli that are not consciously represented. Rossetti and Regnier (1995) explored movements made by healthy controls in conditions known to dissociate the pragmatic and semantic systems in patients and investigated movements made toward targets when recall or verbalization of target location was required. Rossetti and Regnier found larger variability in the pointing scatters when the target position had to be memorized. In Rossetti's study the sighted participant sat before a tactile screen; the starting and end point of the hand for pointing at targets was located at the bottom of the screen (see Figure 3.1). Rossetti's procedure was similar to that of Ittyerah et al. (2007). Pointing movements were executed to visual targets briefly flashed at a random location. Adult participants had to place their index finger on the remembered visual location after variable memory delays (0 and 8 seconds). It was observed that the main axis of the pointing distribution (i.e. the pointing scatter) was mainly oriented along movement direction at 0 second delay indicating the involvement of an egocentric frame of reference and along the direction of the scatter array for 8 seconds delay showing the involvement of an allocentric frame of reference (see Figure 3.2). Movements produced after a delay or when a verbal representation of the goal was built resulted in end point distributions different from those obtained for immediate responses to target presentation. In these conditions end point distributions become contingent upon the context (target array), whereas this is not the case when movements are performed in a reactive mode. These data provide evidence for two different systems that may be responsible for goal directed movements: one system involved in immediate stimulus driven movements, whereas another system taking over after a delay.

Rossetti (1998) demonstrated that similar dissociations could be obtained after a four second delay and that the difference between immediate, stimulus driven pointing and delayed context driven pointing is 4 seconds. A similar difference between pointing movements to immediate and memorized targets has been observed for vision by Goodale, Jakobson and Keillor (1994). This indicates a dissociation between self referent and allocentric codes and may be interpreted along the framework of two main visual processing pathways devoted respectively to action and perceptual identification as is for the visual (Weiskrantz 1986) and somesthetic modalities (Rossetti, Rode & Boisson 1995). This finding is compatible with the observation that reaction times are longer for the conscious perception system than for the motor system (Castiello, Paulignan & Jeannerod 1991) showing differences in the times of the pragmatic and semantic systems used for goal directed movements.

Pointing to memorized targets in accordance to an immediate or delayed instruction has been deemed to be crucial for the manifestation of egocentric and allocentric encoding in reachable space for adult sighted subjects (Rossetti 1998; Rossetti, Gaunet & Thinus-Blanc 1996; Rossetti & Régnier 1995). Rossetti et al. (1996) indicated dissociation in spatial representation following immediate (0 sec.) and delayed pointing (8 sec.) in congenitally blind and sighted adults. Their study revealed that lack of early visual stimulation minimizes differences between representation of objects in space that rely on a contextual frame of reference and that based on immediate self reference. Further, Gaunet and Rossetti (2006) found that the main axes of the pointing distributions differed with the visual status of the subjects for pointing at targets in the absence of vision. For the sighted blindfolded group, the main axis was aligned with the target array for the 8 second delay and with movement direction for 0 second delay. For congenitally blind adults, the orientation of the main axis of the pointing distribution was aligned with movement direction for both delays, whereas the main axis of the late blind group was intermediate to that of the other two groups. Comparing the pointing distributions of congenitally blind and sighted blindfolded children and adults, Gaunet, Ittyerah and Rossetti (2007) observed that the pointing responses of the children confirm that a given system of information processing is involved in immediate stimulus driven movement and that another system takes over after a delay. These findings indicate that the motor representation expressed within a short delay relying on an egocentric reference, and the cognitive representation expressed within a longer delay relying on allocentric reference, is evident at 6 years and are independent of early visual experience. It needs to be emphasized that the endpoint distributions observed in the congenitally blind children have been found to become contingent upon the target array after delay. The comparison between children and adults show that early visual deprivation does not prevent the integration of a global spatial arrangement of target locations in children (Millar 1994) whereas the lack of early visual experience disrupts allocentric pointing during adulthood in the congenitally blind (Gaunet & Rossetti 2006). Thus the ability to dissociate immediate from delayed pointing instructions in congenitally blind children may not be sustained with time during adulthood. This divergence in ability between congenitally blind children and adults may be explained by the long-term experience of the adults in space without vision for processes that rely on an egocentric frame of reference. Furthermore, the effect of the delay found in both groups of children (Gaunet et al. 2007), and among sighted adults (Gaunet & Rossetti 2006) suggests that the dual system relies on convergent information from divergent sense modalities (Rossetti 1998).

3.7 Role of vision in hand actions

Theorists have attributed the accuracy of object perception and reaching, to specific visual pathways involved in the neural control of action (for example Jeannerod 1997a). They argued that visual pathways mediating prehension indicate distinct pathways for visual perception and visual control of movements (Milner & Goodale 1993), and offered suggestions for different visual pathways in the dorsal stream itself which correspond to the hypothesis of various visuo- motor channels one for reaching and another for grasping.

However, the effects of feedback or knowledge of results with sighted children suggest that visual feedback is not always afforded the same privileged status during childhood as in adults, particularly for guiding, reaching and pointing movements (Bard, Hay & Fleury 1990; von Hofsten & Rosblad 1988). The 6–10 years range is of particular interest as it spans a period of development in which the reliance on visual feed-back to monitor and correct reaching has been shown to change. During the development of prehension, 5 and 6 year olds displayed no increase in movement time in the absence of vision, though older children (7–10 years) used visual feed back to improve efficiency (Smyth, Peacock & Katamba 2004; Smyth, Katamba & Peacock 2004). Interestingly, this developmental process has been found to be nonmonotonic. For instance under suppression of visual feed-back during hand movement conditions, 7- to 8-years old children made more spatial errors in pointing tasks than either their older or younger counterparts indicating a greater reliance on vision during this period of development (Hay, Bard, Fleury & Teasdale 1991) (see also Ferrel-Chappus, Hay, Olivier, Bard & Fleury 2002; Hay 1978). It is of interest to know the role of vision and the age related differences for pointing between the congenitally blind and sighted blindfolded children.

A further question of interest is the ability of the hands to represent space for pointing at a sagittal plane. Studies with sighted infants have demonstrated early right hand preferences for discriminating features of objects and left hand preferences for contours of objects (Streri 2002). Streri (2005) observed that at two months of age when prehension- vision coordination is absent the tactile sense is predominantly utilized and touch to vision transfer of information is observed. At five months when reaching and grasping an object is intensive, vision dominates showing successful vision to touch transfer though little touch to vision transfer. Although Streri's tasks involved the manipulation of three dimensional shapes, they resemble pointing and reaching in structure. Furthermore, among blind children that have had no exposure to vision during development, there is a possibility that they rely on convergent information from modalities other than vision (Streri 2005). The hands are well utilized for perceptual and motor functions early in development.

Hand pointing responses are a valuable measure of performance in spatial localization tasks. They indicate different types of spatial representations and the underlying cognitive mechanisms (Paillard 1987, 1991). Whereas goal-directed

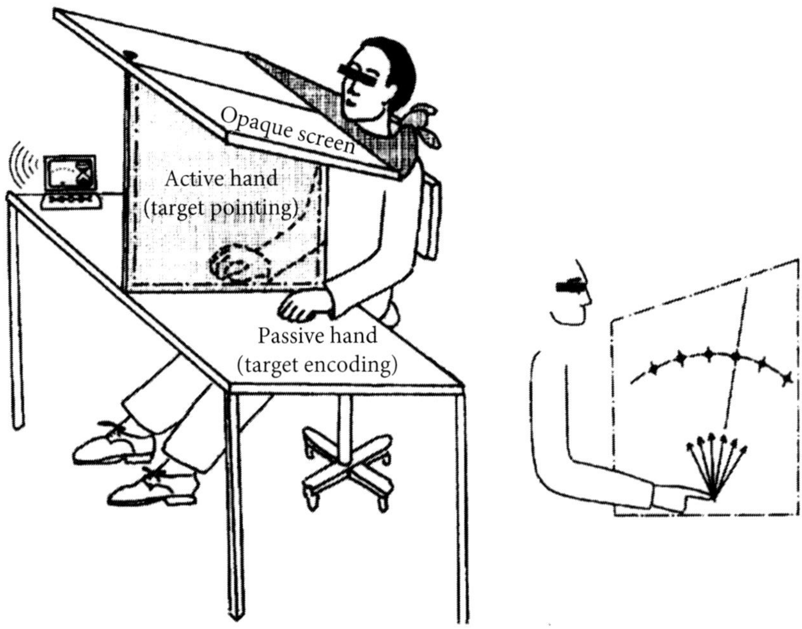

Figure 3.1. The experimental set up

Reproduced with permission of publisher from Ittyerah, Gaunet and Rossetti (2007), Pointing with the left and right hands in congenitally blind children. Brain and Cognition, 64 (2) 170–183. Copyright © Elservier, 2007

pointing has been extensively studied in infants and in adults (Bradshaw & Watt 2002), no study has investigated pointing without vision in children and the properties of spatial representation. Ittyerah, Gaunet and Rossetti (2007) applied the memory guided movement paradigm in congenitally blind and sighted blindfolded children. Our goal was to determine when the effect of delay would appear during development in blind and sighted blindfolded children. It was also of interest to know if pointing accuracy differed between the left and right hands of the blind and sighted children. We (Ittyerah, Gaunet & Rossetti 2007) were interested whether dissociation of immediate and delayed pointing at proprioceptive targets with the left and right hands can be obtained in the total absence of vision in congenitally blind and sighted blind folded children between the ages of 6 and 12 years. The study addressed the processing of space in a task that required differentiation between immediate and memorized representation of targets. Congenitally blind and sighted blindfolded children were required to point at targets on a sagittal screen with their left and right index fingers (see Figure 3.1).

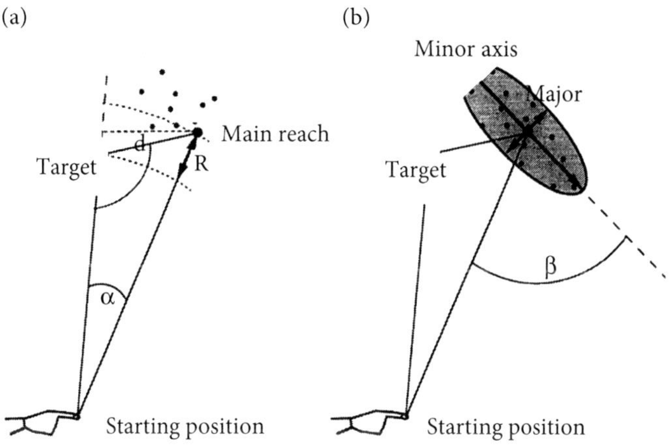

Figure 3.2. Computation of dependent variables
The large dark dot represents the target and the smaller dots represent pointing responses at the target. Left: d, absolute distance error between pointing distribution and target; α, movement direction error of the pointing distribution; R_0, movement amplitude error of the pointing distribution. Right: minor axis length of pointing distribution, smaller dimension of the pointing distribution scatter; major axis length of pointing distribution, greater dimension of the pointing distribution scatter; β, major axis orientation of the scatter
Reproduced with permission of publisher from Ittyerah, Gaunet and Rossetti (2007), Pointing with the left and right hands in congenitally blind children. Brain and Cognition, 64 (2) 170–183. Copyright © Elservier, 2007

Although the non-pointing hand proprioceptively defined the target location in every trial, the emphasis was on the analysis of several parameters of pointing distributions reflecting properties of spatial representation (see Figures 3.2a and 3.2b).

The coordinate of endpoint recording of each individual movement was used to compute several constant, variable and absolute errors. Figure 3.2a, left, describes parameter *d* (absolute distance error), the constant errors R_0 (*movement amplitude errors*) and α (direction amplitude errors); Figure 3.2b, right, describes the constant errors, *x* (errors along the horizontal axis) and *y* (errors along the vertical axis) and the variable errors which are the *scatter surface* (in gray), minor and major axes lengths of the pointing distribution (i.e. smaller and larger dimensions of the pointing distribution, respectively, used to compute the *ratio of lengths of the major–minor axes*) and (orientation of the *major axis of the scatter*). Constant errors are over shooting and undershooting errors with respect to the target locations, whereas variable errors are of three types that depict the surface area of pointing, the lengths of the major and minor axes of the pointing distribution and the orientation of the main axis of the scatter.

We provide below descriptions of these parameters and formulate predictions about the effects of age, group and delay factors on thee parameters.

Absolute error

The absolute distance error or *d* (in millimeters) is the unsigned mean distance between pointing and the target. Although the so-called absolute error (mean distance to the target) is a non-specific parameter affected by both constant and variable factors (Schutz & Roy 1973), its value was computed to allow comparisons with previous studies. Because this parameter is affected by both constant and variable parameters, it is difficult to predict its effect with children.

Constant errors

The output of a sensorimotor transformation may indicate a bias in the internal representation of movement encoding and target position, i.e. the coordinate system used. Constant errors were measured along movement direction (in degrees) and along movement amplitude (R_0 in millimeters), relative to the ideal reach that would hit the target. Angular errors were negative when the reach was closer to the body than the aimed target (undershooting in direction) and amplitude errors were negative when movement distance was shorter than the ideal reach (undershooting in amplitude).

The constant error along the horizontal (or x in millimeters) and along the vertical (or y in millimeters) axes is the mean of the algebraic errors in pointing movements to the centre of the target. Errors along the horizontal axis were negative when pointing projections on the x-axis were below the projection of the ideal reach (undershooting along horizontal); errors along the vertical axis were negative when pointing projections on the y-axis were under the projection of the ideal reach (undershooting along vertical). The covariation of errors along the horizontal axis and movement direction and errors along the vertical axis and movement amplitude is predicted.

Variable errors

The 2D variable errors were assessed by the computation of confidence ellipse parameters. Confidence ellipses (i.e. two-dimensional mean ±1.96 SE of pointing responses) were derived for each subject, each delay and each target position. They were basically characterized by the length of the two orthogonal axes computed by orthogonal regressions between x and y coordinates of each scatter endpoint. When comparisons are made between two groups of ellipses, an increase of the length of the major axis implies that the pointing distribution is more elongated along the main axis of the scatter, whereas an increase of the minor axis of the ellipse implies that more pointing variability is observed along the direction orthogonal to the main axis of the scatter. The ratio between these two measures specifically describes the elongation of the ellipse. In addition, the total ellipse surface (minor × major × π) provides a precise estimate of the two-dimensional variability of the pointing performance, which depicts the surface of the probability zone (95%) where the endpoint mean is likely to be. The *scatter surface* (in mm^2) provides an estimate of the global pointing variability, and is thus expected to increase with the longer memory delay (Rossetti et al. 1996). These parameters allow only for a quantitative assessment of the pointing variability, whereby systematic trends within the global variability are masked, that is, the scatter may vary in shape. Two additional parameters were computed in order to describe the qualitative spatial (and not the quantitative) aspects of each pointing scatter. First, the ratio of the two axes lengths provides a description of the ellipse shape, ranging from a circle (ratio = 1) to a more or less thin distribution around the main axis. According to previous studies, no effects of group or delay are expected to affect the ratio (Rossetti et al. 1996). Second, the orientation of the major axis of the ellipse (β) was analyzed in polar coordinates centered on the starting location. To this aim the orientation of the main axis of each ellipse was computed with respect to the average pointing direction. The angle between the mean movement direction (from the starting position to the average pointing location) and the major axis of the scatter of each target was calculated. The resulting angle could range between -360^0 and $+360^0$. As each β could appear as either a positive or a negative value, these angles were then converted to modulo π (i.e. a constant) within a half trigonometric circle. According to previous studies, values were expressed in angles ranging from 45^0 to 225^0 (a range of 180^0) in order to obtain unimodal distributions of β. As a matter of fact, previous studies (Rossetti et al. 1996), (Rossetti & Pisella 2002), (Rossetti & Regnier 1995) and (Rossetti et al. 2000) have shown that values of β are typically distributed around 90^0 and 180^0 within the circumscribed range. In addition the chosen half trigonometric circle was such that these values lie symmetrically within this interval, i.e. 45^0 from the edge rather than at 0^0. The interpretation of this parameter is straight-forward. When β values are close to 180^0, they indicate that the scatter is elongated in the direction of the pointing movement. When β is close to 90^0, it indicates that the scatter elongation is orthogonal to the pointing direction. For example, if an arc-shaped target array centred on the starting location is used, this implies that the pointing distributions are primarily aligned with the target array if β values are close to 90^0. Note that this parameter was neither a measure of pointing variability nor a measure of constant pointing error. Instead, it provided a cue to the spatial frame of reference used by the subjects to perform the task.

To show the strength of local distortion (or the involvement of ego- or allocentric representations) the analysis of both the ratio of the two main axes lengths, and the orientation of the main axis, have to be considered. For β to be meaningful the ratio of the two axes should not be equal to 1, otherwise the distribution would be too circular for its orientation to mean anything reliable. According to previous findings in adults (Rossetti et al. 1996; Rossetti et al. 2000 and (Rossetti & Regnier 1995), this orientation should be affected by the delay in blindfolded sighted children. In theory, β should be aligned with movement direction at 0-s delay as is observed for visual targets (e.g. Rossetti 1998; Rossetti et al. 2000). Our most crucial test will be to analyze the pointing performance in the delayed condition. As suggested by Newcombe and Huttenlocher (2000), if alloocentric representation develops in children and the blind like the sighted children are able to perform in allocentric space (Millar 1994), then β can be expected to become aligned with the target array in the delayed condition for both groups of children.

It was of interest to note that the endpoint distributions observed in congenitally blind children were found to become contingent upon the target array during the longer delay like that of the sighted blindfolded children, for the right hand pointing responses. Though the absolute error was greater for delayed pointing, there was an effect of development for both hands. The blind children occupied lesser surface area of pointing than the sighted, showing better accuracy of pointing at the targets. This finding reveals a steady tactile development in the blind during infancy, and suggests the development of some parts of the brain that may undergo significant postnatal development (Streri 2005).

Of interest is the unexpected difference in pointing accuracy ('d' and surface area) with each hand. The left hand in both groups of children was greater in absolute error and surface area of pointing. The left hand also differed from the right hand in its orientation to the major axis of the scatter of pointing responses. This indicates that the hands tend to differ such that the left hand is more egocentrically oriented while the right hand is relatively context oriented (see Figure 3.3).

Figure 3.3 shows the orientation of the main axis for pointing at targets with the left and right hands in both congenitally blind and blindfolded sighted children. The egocentric orientation of the left hand seemed to have increased the surface area of pointing and absolute error in the left hand. This may explain why in some studies fine motor skills that are context oriented, are better performed by the right hand (Healey, Lederman & Geschwind 1986; Steinhuis & Bryden 1989, 1990; Ittyerah 1996; Streri 2002), whereas less lateralized tasks such as lifting objects (Steinhuis & Bryden 1989, 1990) that show actions of strength than accuracy are better performed with the left hand (Healey et al. 1986; Peters 1990c; Ittyerah 1996).

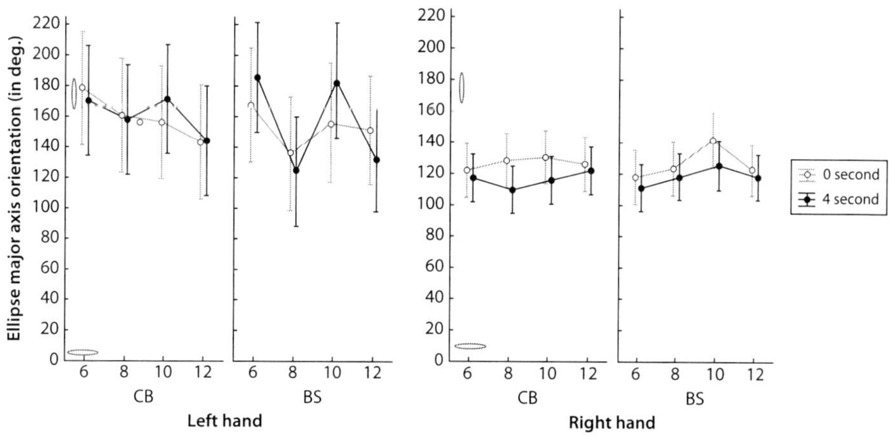

Figure 3.3. Main axis of the scatter for the left and right hands for the blind and sighted blindfolded children at 0 and 4 seconds delay

Reproduced with permission of publisher from Ittyerah, Gaunet and Rossetti (2007), Pointing with the left and right hands in congenitally blind children. Brain and Cognition, 64 (2) 170–183. Copyright © Elservier, 2007

The pointing distributions of the two groups of children in the absence of vision indicate that pointing is not affected by the visual status but for the finding that the congenitally blind group have a smaller surface area of pointing than the blindfolded sighted children (see Figure 3.4). Therefore by not restricting the spatial experience only to vision, a more consistent representation (global pointing variability measured by the scatter surface) of the target location is obtained as a consequence of converging sensory experience (Rossetti 1998). Overall these results indicate that vision is not necessary for a spatial framework since consistent effects of development have been found in blind children as in their blindfolded sighted peers. The present findings suggest that congenital blindness may not be detrimental to the development of spatial representation neither self referential nor contextual, since there is evidence that people who are totally blind from birth can perform as well (Millar 1994) or more proficiently (Hollins 1986) than the sighted on spatial tasks. Casteillo, Bennet and Mucignat (1993) for example observed that experience of vision is not necessary for the coordination or patterning of the basic reach to grasp movement in blind adults. Evidence (Millar & Al-Attar 2005) indicates that vision improves performance in a haptic spatial task only in so far as it adds cues that are potentially relevant to spatial discrimination and reference. The performance of sighted blindfolded subjects with diffuse vision was no better than performance without vision. Therefore vision does not affect haptic processing if it does not add task relevant information.

Although much evidence about the role of vision in hand use reveals a coordination between eye and hand for reaching or pointing at objects (von Hofsten 1982; Jeannerod 1997a, 1997b; Johansson, Westling, Balkstron & Flanagan 2001, etc), studies indicate that processes underlying pointing are early and that infants do not need to see their hand to reach and contact a toy (Clifton, Muir, Ashmead & Clarkson 1993; McCarty, Clifton, Ashmead, Lee & Goubet 2001). These empirical findings indicate that visual guidance of the hand is not necessary to establish object contact about the onset of reaching and therefore proprioception is sufficient for pointing very early in development. Furthermore, before 7-years of age, perception of positions is predominantly based upon proprioception (Hay 1978). The findings indicate that the delay effect was similarly pronounced in the blind as well as the sighted blindfolded children and this suggests the involvement of the two processing systems (Rossetti 1998) that are not dependent on early visual experience. We also found that at 4 seconds delay, the scatter surface was larger than that at 0 second delay for both groups of children confirming that a delay between memorization and the output deteriorates the concentration of the pointing distribution (see Figure 3.4).

Our (Ittyerah, Gaunet & Rossetti 2007) findings confirm that a given system of information processing is involved only in immediate stimulus driven movement and that another system takes over after a short delay. Consequently, these pathways are also related to egocentric and allocentric encoding (Goodale & Milner 1992; Milner & Goodale 1993; Rosseti 1998). According to our results, the former system devoted to action is self referent and the latter system (perceptual identification) is allocentric and performs according to a global representation of the spatial design involved in the task. Our (Ittyerah et al. 2007) study shows that the visuo-motor (pragmatic) representations expressed within a longer delay relying on allocentric representation was evident at 6 years and was independent of early visual experience. The study provides crucial descriptive data and a theoretical perspective concerning the development of cognitive processes in pointing without vision. Both movement (self referential) and target array (contextual) frames of reference are present in young children. The dissociation of spatial encoding processes according to memory delay in young children, as well as in children totally congenitally blind (independent of visual experience) suggests that the double system relies on convergent information from different sense modalities (Rossetti 1998).

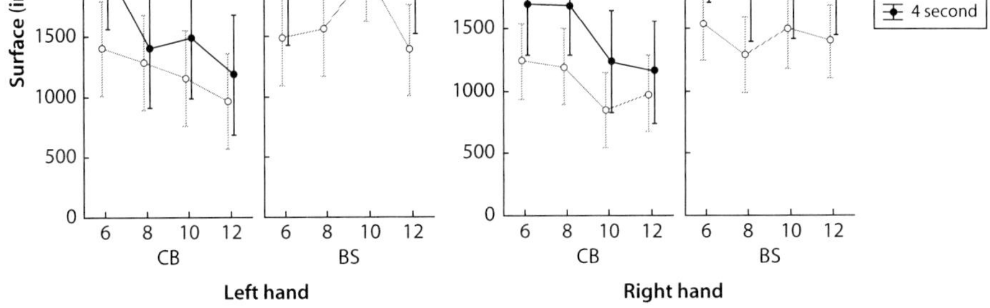

Figure 3.4. Pointing scatter surfaces of the blind and sighted blindfolded children at 0 delay and 4 seconds delay

Reproduced with permission of publisher from Ittyerah, Gaunet and Rossetti (2007), Pointing with the left and right hands in congenitally blind children. Brain and Cognition, 64 (2) 170–183.

Furthermore there is a difference in the strength of the frame of reference for boys in both blind and sighted groups, with a larger scatter elongation than the girls, suggesting a gender advantage for spatial representation. Studies continue to show that boys have an advantage for spatial tasks (Lowe et al. 2003) and this is evident during adulthood for mental spatial rotation tasks on paper (Parson, Larson, Kratz & Thiebeux 2004) and spatial memory (Postman, Jager, Kensels & Koppenschaar 2004). We conclude that since the boys were able to distribute their responses along an arc particularly for the longer delay with larger scatter ratios, they are more contextually oriented than the girls regardless of their visual status. The results also suggest that the perceptual identification pathway enables a global representation of the spatial design involved in the task. Such an information processing system would allow the children to distribute their pointing along the target arc within an external frame of reference. Thus contextual, cognitive factors govern haptic cognition and they differ from implicit, immediate factors such as immediate action toward the target.

Even if speculations about the ability to point among the sighted children may be attributed to their temporary blindfold conditions that were expected to hamper the performance of the preferred hand, there was no reason to expect a similar decline among the blind children who were also right handed. Therefore though vision may provide external references for the sighted, the blind were found to use self- reference cues during performance and visuo-spatial proficiency has been found to improve under blind conditions as well (Liben 1988; Millar 1994). In completely blindfolded conditions there is usually little reliable information about external frames for coding small scale displays. Body- centred frames are reliable and self referent coding can be used (Stelmach & Larish 1980). When subjects are stationary in blindfold conditions, information is restricted to personal space that is to spatial locations within the arm's reach without moving bodily to another place. Such conditions are of particular interest in studying both short- and long-term effects of modes of perception on coding.

Undoubtedly, visual information increases precision in performance, though the criticality of vision as an omnipotent pervasive factor in hand preference or skill is not evident. Therefore efforts should be directed toward studying the role of the hands in task performance rather than the role of vision in task performance. The findings of hand ability indicate that there is little or no relationship between the preferred hand and actions of skill. The left and right hands seem as able as each other in tasks that assess most manual skills (Millar 1997; Millar 2008; Ittyerah 1993, 2000, 2009) and these effects have been consistently found across large samples over varying ages of development in the total absence of vision and in sighted blindfolded conditions.

3.8 The hand as a frame of reference

The previous section discussed how instructions to reach targets, either immediate or delayed can alter spatial frames of reference. The present section will discuss the role of the hand as a frame of reference in haptic parallelity tests. Haptic parallelity tests indicated that while participants matched the orientation of a test bar as being parallel to a reference bar, large systematic deviations from physical parallelity existed in the horizontal and midsagittal planes (Kappers 2002, 2003, 2004, 2005). Subsequently Volcic, Kappers and Koenderink (2007) tested sighted blindfolded participants to know whether systematic deviations from what is physically parallel can be detected on the frontoparallel plane. The participants were required to perform a bimanual parallelity task. The participant stood before a board on which different locations of the reference and test bars were distributed symmetrically on the left and right side of a board. The left hand touched the bar on the left side and the right hand touched the bar on the right side. The participants were instructed to rotate the test bar in such as way that they felt it to be parallel to the reference bar.

Results showed large deviations in the parallelity task for the frontoparallel plane as in the horizontal and midsaggital planes of previous studies (Kappers 2005). There was good evidence of the involvement of intermediate frames of reference from the fact that the degree of hand/arm rotation correlated with the amount of deviation (Kappers 2005). Findings supported predictions that participants with high average deviations would be prone to rely on an egocentric frame of reference and thus display a stronger correlation between hand/arm orientation differences and deviations. Participants with low average deviations that may have relied on an allocentric frame of reference did not exhibit any dependence either on hand orientation differences or on relative distance between bars. Though the usual oblique effect was found for participants with small deviations between the reference and test bars, the reverse oblique effect where the error is greater for cardinal than oblique positions was observed for participants with large deviations between the reference and test bars. Thus the magnitude of deviation is affected by the degree to which the egocentric or allocentric frame of reference is dominant.

Just as two dimensional haptic processing was prone to systematic deviations evidence showed similar systematic deviations were found for three dimensional haptic parallelity tasks. Volcic and Kappers (2008) tested sighted blindfolded participants unimanually on a 3-D parallelity task. The stimuli were rotatable aluminum bars that were placed on top of poles distributed on a table from left to right and near to far. The 3 D orientation of the bar was defined by tilt and slant. Bars were used both as test and reference bars. The subject sat before the experimental set up with her/his body midline aligned to the mid-point of the set up. At this position all bars on the poles were within reach without bending or displacing the body. Participants were permitted to use their fingers and palms to touch each bar though they could approach each bar only from above so that they did not touch the sides or the poles on which the bars rested. The orientation of the reference bar was fixed and the right hand of the participant was placed on the bar. The participants were instructed to rotate all eight test bars in a way they felt all the bars were parallel to the reference bar. Further, the hand orientation of each subject was measured for each bar position in order to examine its influence during the parallelity task. It was assumed that hand orientation corresponded to the orientation of the egocentric reference frame at each bar location. The orientation of the hand was specified by the tilt and slant angles of the bar. The tilt angle referred to the up-down orientation of the hand and the slant angle referred to the left-right orientation. The hand orientation of each subject was measured for left-right and the far-near conditions.

Results showed that participants deviated from veridicality when required to construct a field of parallel bars in three dimensional space. The systematic pattern of deviations was captured by the hand centred weighted average model that presupposed a biasing or interfering impact of an egocentric reference frame fixed to the hand on the allocentric frame of reference.

Although subjects differed in their average weighting factors, their range was very small. On a continuum of a reference plane fixed to space and to the hand, performance shifted on average by 19.6% from the allocentric reference frame to the one fixed to the hand. This finding was not different from the 23.8% shift toward egocentric reference frame reported by Kappers (2007) in a two dimensional parallelity task. Thus the hand is used as a reliable frame of egocentric reference in space.

3.9 Sensorimotor interface

Neural representations of changing somesthetic experiences received or sought by the hand exist in the somatic afferent pathways of the brain and somatic sensory areas of the cortex. Stimulus parameters such as neural activity and behavioural performances can be observed and measured. The important event is the flow of activity over somatic afferent system to the posterior parietal lobe and parietal motor areas of the frontal lobe that link sensory to motor actions at the cortical level (Mountcastle 2005).

Studies of world war I gun shot wounds by Head and Holmes in Britain and Kleist and Goldstein in Germany established that the body is represented in the post central gyrus in a regular pattern with the foot most medial and face most central and tactile sensitivity appears to be localized closer to the central sulcus. The closer a lesion to the central sulcus, the more likely it is to produce defects in the primary aspects of touch. The post central gyrus generates somatic sensory function such as the recognition of form by the hand, knowledge of position and movement of limbs in space, the generation of an internal schema of the body form and the generation of the movements of the arm and hand directed at objects (Mountcastle 2005). The posterior parietal system generates the transcortical projections linking the parietal and frontal lobes that initiate and govern projected movements (Mountcastle 2005).

The posterior parietal cortex is considered to be a sensory motor interface in the planning and control of visually guided movements (Buneo & Anderson 2006). The posterior parietal cortex (PPC) is known to convert sensory information to motor commands and integrate sensory input with previous and ongoing sensory commands to maintain a continuous estimate of arm state that can be used to update present and subsequent movement plans in an eye centred frame of reference.

Mountcastle (2005) summarized from the findings that the left parietal lobe governs motor attention throughout visual space, as well as spatial attention in the right visual field. The right parietal lobe governs spatial attention throughout visual space as well as motor attention in the left visual field. Transitions from sensation to action are initiated by activity in the parietal frontal cortical systems in which integrative and planning operations are generated and expressed in the patterns of discharge of their constituent neurons, such as in the parietal reach region for the movement of the hand and arm to the target, or in the anterior intra parietal area for the formation of hand movements to match the 3 D structure of the object.

The experience of touch may go beyond the hand to incorporate the whole body. This is usually an unconscious process that requires the integration of multiple sensory inputs. The experience of touch can also involve the embodiment of habitually used tools, prosthesis or rubber limbs that effectively extend or displace the influence of body parts. Information about body in space is an integration of afferent sensory information from the retina, somesthesis, proprioception, audition, for example together with efferent information relating to the movement of the body in space (Stein 1992; Ventre-Dominey, Nighoghossian & Denise 2003). Regions within the posterior parietal cortex integrate these multisensory inputs and outputs (Bonda, Petrides, Frey & Evans 1995). When a movement command is generated, an efference copy of the movement is sent to the parietal cortex toward the superior parietal lobe (Wolpert, Goodbody & Husain 1998) to generate expected sensory outcomes (McGonigle, Hanninen, Salenius, Hari, Frackowiak &

Frith 2002). During tool use by the hand the somatosensory cortex retains a schema of the hand in action (Schaefer, Rothemund, Heinze & Rotte 2004). Tool use is lateralized to the left hemisphere and becomes represented as an extent of the hand within the body schema (Lewis 2006).

When a tool is used to reach a distant object, this extended motor ability is followed by changes in particular neural networks that have an updated map of the body posture. These changes are compatible with the notion that the tool is a part of the body schema (Marivita & Iriki 2004) as if the hand (effector) were elongated to the tip of the tool. Evidence (Colby 1998; Graziano & Gross 1998) indicates that multiple fronto parietal networks integrate information from discrete regions of the body surface and extend space in a functionally relevant way to specific actions performed by different body parts. Data from monkeys, normal humans and neglect patients support the idea that when tools are held in the hand, they enable extent of reaching space that can become incorporated into a plastic neural representation of the body (Head & Holmes 1911). Holmes, Calvert and Spence (2004) indicated that only the tip of the tool becomes incorporated into the body schema. Though active tool use alters the neural representation of peripersonal space, such that a short period of tool use can lead to an expansion of spatial representations to last a few minutes after tool use, multisensory interactions may change after each trial and is dependent upon the predictability of the next movement (Holmes, Calvert & Spence 2007). Certain effects of hand held tools on multisensory interactions may be motor and may include attention and movement preparation components. Therefore tools serve as spatial attention and motor cues to highlight certain regions of space relevant to the task. It is the tip of the tool and not the middle portions that are represented during multisensory interaction to be incorporated into the body schema (Holmes et al. 2004).

Active tool use can plastically alter the nature of crossmodal extinction (Farne & Ladavas 2000) and decrease crossmodal extinction (Maravita, Clarke, Hussain & Driver 2002) in patients with unilateral lesions. Such patients typically fail to report tactile stimuli on the contralesional hand (that is the hand in the opposite hemispace of the lesioned hemisphere), when these stimuli are presented together with competing visual stimuli near the ipsilesional hand (the hand on the same side as the lesioned hemisphere) (Bender 1945; Driver et al. 1997). Maravita et al. (2002) speculated that active tool use in these patients may extend crossmodal interactions between visual stimuli at the tip of the tool and tactile stimuli on the hand wielding the tool. By this intervention they expected to overcome the competitive effects of spatial attention (Duncan 1984) demanded by ipsilesional stimuli in the presence of contralesional stimuli that get undetected by such patients. Prolonged tool use lasting for 10–20 minutes with the contralesional hand in ipsilesional space reduced crossmodal extinction considerably and its effects lasted for over an hour. Such effects of training with tools may occur laterally and across the midline to reduce pathological crossmodal extinctions and such lasting effects of tool use may be of interest for rehabilitation.

3.10 Handedness and species differentiation

Despite neural evidence representing hand actions in the brain, the origin of hand preference is yet inconclusive. According to Annett (2002), in the course of human evolution small changes in the genome could have allowed changes in favour of right handedness. Amongst the nonhuman species, the chances of left and right paw preferences are nearly equal at the population level. Amongst humans the greater number of right hand preference (66 percent according to the right shift theory) is attributed to a change in a single allele at one locus that accounts for the RS+ gene. Presently, the mechanism of the RS gene is unknown, but it is inferred to weaken the role of the right hemisphere in early learning, so that the infant is more likely to use the left side than the right, in early phases of speech and language learning. However, the RS+ gene is neither necessary nor sufficient for speech because the gene is not present in 20% of the population (Annett 2002). Nonetheless, although the RS gene is not a prerequisite for speech, the promotion of left hemisphere lateralization is likely to be an important aid to the efficacy of speech and language learning. The gene assists the development of language skills at the risk of some imbalance of cerebral function due to loss of right hemisphere capacity. Annett explains that outstanding achievement in any area that requires maximum efficiency on both sides of the brain and body (as in surgery, playing musical instruments) may be less likely in those with RS++ genotype. According to Annett it is important to recognize that handedness is not the cause of any language problem or a special skill. Handedness occurs by chance and is a consequence of the RS+ gene. Those who possess the gene are more right side lateralized than those with a RS– gene, or the absence of an RS gene. Annett clarified that the RS+ gene evolved to promote speech and not language.

It has been observed that people with sex chromosome anomalies have different patterns of relative ability for verbal and nonverbal skills. Turner syndrome females who lack one X chromosome tend to have average verbal abilities but below average nonverbal, visuospatial abilities. Individuals with an extra X chromosome, such as XXX females, or XXY males, tend to have average nonverbal abilities, but weaker verbal ability. These differences led Crow (1998) to suggest that a gene relevant for cerebral dominance is located on the sex chromosome. According to Crow (1998, 2000), a single gene for cerebral dominance was a critical factor for human speciation. For Annett (2002), sex differences were not caused because the RS+ gene was located on the sex chromosome. The expression of the gene is modified by factors that influence growth and the rate of early growth differences between the sexes, that is, the rate of fetal growth is faster for females than males and faster for single born than twins. Annett believed that gender characteristics such as better language skills in women and better physical prowess in males complement each other as a consequence of maturation and not chromosomal differences. However, Annett has not accounted for the asymmetries in the human skull such as the greater left than the right occipital width and a greater right than left frontal width, which Crow considers to be an essential determinant of human speciation. Crow asserts that no homologue of the cerebral torque from right frontal to left occipital lobes has been established in nonhuman species. Thus the null hypothesis that the torque is the *Homo sapiens* specific neural basis of language has not been disproved. Crow claims that schizophrenia is a language disorder and individuals suffering from schizophrenia are less lateralized. One suggestion is that schizophrenia represents a component of the genetic variation associated with the establishment of dominance in one or other cerebral hemisphere, a mechanism that has been crucial in the evolution of language. Crow, Done and Sacker (1996) found that pre-schizophrenic children were more likely to be rated by their mothers as ambidextrous. At the ages of seven and eleven they were less strongly right handed than their peers in a cohort population on a test of relative hand skill such as speed of checking squares on a sheet of paper with the left and right hands and picking matches with the left and right hands. Therefore delay in establishing dominance in one hemisphere could be the critical factor that predisposes to schizophrenia.

In a follow up study, Crow, Crow, Done and Leask (1998) tested population variation in handedness in a large sample of children. They assessed relative hand skill in a square checking task as a predictor of verbal, non verbal, mathematical and reading comprehension ability at the age of 11 years. The most substantial deficits in ability were seen close to the point of equal hand skill, indicating hemispheric indecision. For verbal ability, females performed better than males, but the relationship to relative hand skill was quite similar for the two sexes; for reading comprehension, males close to the point of equal hand skill showed greater impairments than females. There was a relationship between relative hand skill and cognitive ability. The failure to establish dominance in one or the other hemisphere was problematic. Those that failed to show a definite hand preference at least by the age of eleven years were globally delayed in their development relative to their peers. These deficits were consistent with the concept of Orton (1937) and Zangwill (1967) that cerebral ambilaterality is associated with a delay in development. More recently Leask and Crow (1997) found a decrease of verbal skill in those who were close to the left is equal to right hand performance (L=R). Children premorbid for schizophrenia were less lateralized and their verbal skill was lower than predicted by their hand skill, with a similar trend in children premorbid for affective psychosis. Thus pre-psychotic children deviate from the general population in the trajectory of lateralization of words. The findings were consistent with the concept that in psychosis at some critical stage in development there is a failure of lateralization of the components of language. Therefore handedness constitutes an observable behavioural lateralization and its significance depends on the degree to which it is associated with lateralization of higher cognitive tasks including language.

Chance and Crow (2007) have argued that handedness, brain anatomy and cognition in great apes, extinct hominids and modern humans are evidence that suggest cerebral lateralization underlies the human capacity for language. Though chimpanzees do not show consistent population level handedness, directional handedness on a population is present and universal in modern human sapiens. Chance and Crow emphasized that in the anatomical structure of the brain, there are components of macroscopic asymmetry that are dissociable: (1) The petalia or the protrusion of the right frontal and left posterior poles relative to their counterparts in the opposite hemisphere have an older evolutionary origin and are not related to language and (2), a volume asymmetry that seems to be *Homosapiens* specific and associated with the asymmetry of cell organization in human language cortex, not found among apes. Crow (1998, 2000) proposed a four- quadrant model reflecting the fact that the cerebral torque constitutes a bias across the anteroposterior axis, such that the left and right prefrontal cortices are asymmetric and have the reverse asymmetry of posterior occipital-parieto-temporal cortex. Crow depicts the sensory areas as being in the anterior regions of the brain and the motor areas in the posterior regions. The right half of the brain is considered to be abstract with abstract conceptual areas in the posterior sensory regions and abstract intentional areas in the anterior motor regions. The left half of the brain is considered to constitute the signifiers (De Saussure 1916) with Wernicke's area being in the posterior sensory region and Broca's area being in the anterior motor region.

The encephalization quotient (EQ) is considered to be a brain measure that has evolutionary significance. The EQ represents the degree to which brain size in a species exceeds or falls below the brain size of typical mammals of similar body size. EQ correlates with feeding behaviour in various mammalian taxa. Primates have higher EQ than insectivores or foliovores (Jerison 1973; Gibson, Rumbaugh and Beran et al. 2001). *Homosapiens* have higher EQ than apes and it has been argued that high EQ is associated with greater intelligence. Increased brain size and enlarged cortical surface will occur as a result of minicolumn proliferation and expansion (Chance, Casanova, Switala & Crow 2006). Chance and Crow (2007) indicated that minicolumn asymmetries are present in humans unlike apes and minicolumn number correlates with callosal axon number and reflects the sex difference. Chance et al. (2006) found minicolumn in region planum temporale more asymmetrical in men than women, although mean planum temporale volume relative to brain weight was 11% larger in women than men. Witelson, Glezer and Kigar (1995) found that women have greater density of neurons in posterior planum temporale. Sex differences exist in the development of lateralized functions including verbal skills that are associated with handedness. Sex differences are also found in the connections between the cerebral hemispheres and in the spacing of minicolumns and their constituent cells (Chance et al. 2006). The presence of sex differences in humans and their absence among chimpanzees or primates seem to suggest that sexual selection has played a role in the evolution of the lateralized human phenotype. Thus lateralization is crucial to both language and handedness in *Homosapiens.*

In summary evidence indicates that children premorbid for schizophrenia are less lateralized and their verbal skill is lower than predicted by their hand skill. The findings are consistent with the concept that in psychosis at some critical stage in development there is a failure of lateralization of the components of language. Handedness constitutes an observable behavioural lateralization and its significance depends on the degree to which it is associated with lateralization of higher cognitive tasks including language. Gender differences in humans exist in the development of lateralized functions including verbal skills that are associated with handedness.

The present section on hand and brain has reviewed the role of the cerebral hemispheres in hand actions. Even if hand advantages have been observed for task specific characteristics, the ability of each hand in task performance cannot be ignored. The next section contains evidence from touch and haptics. To know the characteristics of an object such as its shape, size or texture the movement of the hands and its fingers over the object is mandatory. This reflects the cerebral processes that are present in haptic or active touch. It is now known that the right hand controlled by the left hemisphere is better at movement processes, whereas the left hand controlled by the right hemisphere has an advantage in detecting spatial characteristic of the stimulus/object. Therefore the complementary role of both hands must be appreciated for touch and cognition.

CHAPTER 4

Tactile cognition

4.1 Sensitivity

A complete understanding of tactile processes cannot be obtained from studying either visually informed or sighted blind fold individuals in tactile nor cross-modal matching conditions. When subjects are temporarily blindfolded, they are at a disadvantage of an immediate loss of sight, not having the practice of using sources of information that they may in their sighted existence totally ignore. There is also the possibility of visualizing or using spatial cues of reference from the environment to enable stimulus comparisons. Therefore studies with blind subjects are important if experiences devoid of vision have to be understood. Though all of the human body is sensitive to touch, the present focus will be on the role of the hands in the acquisition of information.

In touching an object the stimulus may arouse cutaneous activity that may be confined to only the skin receptors, thus arousing specific sensations that are localized in the skin. This is considered to be relatively passive to sensations that are aroused when the skin is in concert with muscles and joints while exploring an object that consequently provides a percept of the stimulus. It is an active process that is determined by informative relations between stimuli, rather than singular sensations of pressure or warmth. Information about shapes or the surface texture of objects or their arrangements in space must be acquired by active touch that involves the movement of the fingers and palms of the hand. Katz (1989) described the rich world of information provided by touch and showed that the texture of a solid substance could be perceived by touch if there was relative motion between the surfaces of the skin. Touch modality conveys information about a number of different although overlapping skin sensations. Inputs can arise from vibrations or temporally spaced pulses, from pressure that gives impressions of hardness or softness, from patterns of roughness and smoothness, and dry and wet textures. Heat and cold is also sensed. The impressions can also be used to identify objects from extended surfaces and the spatial relations between them. Touch is not a slow and impoverished modality (Millar 1988) as compared to vision. Katz (1989) and Revesz (1950) argued that the hand is a kind of sense organ as different from the skin of the hand. Their subsequent investigation in the perceptual capacities of the touch modality led Revesz to propose an unrecognized mode of experience called haptics that involves more than the classical modalities of touch and vision.

The sense modality of touch is the earliest to develop during the fetal period, followed by audition and lastly vision. Yet touch is a relatively less studied modality though the sensory and cognitive effects during development are all pervasive and lasting. Any sensation of touch on a region of the body surface is aroused by the movement of the skin or tactile structure in contact with an object (Weber 1978). The contact of the object and the skin either compresses or expands the skin and consequently stimulates the area of contact. The skin is either compressed by movement of the object touching the skin it resists, or by movement in the area beneath the skin, which the touched object resists.

A series of fascinating psychophysical experiments by Weber (1978 translated version) in his classic treatise 'The Sense of Touch' revealed varied degrees of sensitivity on different parts of the body. The most sensitive parts of the body are the mouth regions such as the lips and the tongue and the fingertips and palms of the hand. Indeed Weber also emphasized in the second half of his treatise, the dependence of the senses on each other so as to attain convergent sources of information. The explanation of locating pain arising from a toothache to be in the tooth than in the nerves of the gum suggests, is that we rely on indirect sources of stimulation in arriving at conclusions of the origin of a particular sensation. It also suggests intrinsic differences between the different sense modalities. For example Jastrow's (1889) findings of perceptual differences between the eye, the hand and the moving arm in length matching indicates, that although convergence of information between the modalities is more informative than information from only a single modality, the nature and importance of modality specific information cannot be ignored.

For convenience, the tactual system is largely divided into the cutaneous (tactile) and haptic subsystems (Loomis & Lederman 1986). The cutaneous system uses sensory inputs obtained from receptors embedded in the skin. The haptic system uses both cutaneous and kinesthetic information from receptors in muscles, tendons and joints. Traditionally most studies of cutaneous sensitivity consist of psychophysical experiments (Weber 1978) that measure threshold on different parts of the body. These classic findings of tactile sensitivity (Weber 1978) are documented as fundamental contributions for further work. The scientific discoveries of Weber that followed from his experiments on the sense of touch indicated the relevance of tactile acuity as an important measure in the processes of sensation and perception. According to Weber there are two sorts of sensations that are aroused in the sense organ, one is pressure and the other is warmth or cold. In testing for acuity in different parts of the body, Weber used an instrument like an aesthesiometer, known as 'stangenzirkel' in German, or a beam compass with two adjustable sliding points on the beam, so that the two points of contact on the skin can be increased or decreased. Weber found that the two point threshold differs in different parts of the body with the fingers and the mouth being the most sensitive than the rest of the body. Sensitivity is greater when the two points of the beam are placed transversely across the forearm than when it is placed longitudinally. Weber hypothesized that the increase or decrease in sensitivity in different parts of the body is an indication of the more number of nerve fibres that terminate in these sensitive parts than in parts with a dull sense. The head possesses better tactile acuity than the trunk and other parts. Parts are less sensitive the further they are from the extremities of the mouth that have the best tactile acuity and the extremities of the limbs are more sensitive than its other parts.

4.2 Perceiving weights and temperatures

Weber (1978) demonstrated that the weight of objects is perceived in two ways. Firstly, by the touch sense that resides in the skin and secondly, by the sense of voluntary muscles by which we recognize the effect involved in overcoming the resistance of weights and other objects by lifting them. Weights arouse a stronger pressure sensation in the more sensitive areas of the body. When the same weights are placed on different areas, the sensation is similar, but not quite equal if the weight lies on a larger than a smaller skin surface. The more particles of skin a weight rests upon, the less any particle is pressed and vice versa. A difference is perceived between heavier weights and between lighter weights.

Weber (1978) observed that temperature differences between objects are best perceived in areas where the tactile acuity and weight discrimination is best. For example, Weber observed that if one were to immerse the hands together in a flask of hot or cold water, with the ulnar edge downwards and the radial edge upwards and the two hands slightly apart, initially the back of the hands with the thinner cuticle will feel the temperature first. But because the palms possess fine sensibility owing to its greater number of nerves, the heat and cold gradually overcome the obstacle of the thicker cuticle and have a stronger and a more consistent effect in the palm. Thus the sense of temperature and the sense of touch belong to the same sense and therefore it is likely that the same nerves perceive pressure and heat. Weber explained that a smaller degree of heat received from a larger skin surface causes the same sensation as a larger degree of heat stimulating a smaller skin surface. When a larger skin surface is touched by warm objects we recognize a difference in its temperature more clearly than when a smaller surface comes into contact with the same object. Weights and temperatures are perceived more accurately if they touch the skin successively than simultaneously. As a consequence of his experiments with weights, Weber arrived at a constant fraction that enabled the judgment of differences in the perception of weights. According to Weber, when perceiving differences between weights, the ratio of the differences between the weights is a constant proportion of the magnitude of the original stimulus. For example if two weights one weighing five grams and another weighing ten grams were to be compared for heaviness, the ratio of the weights, that is two, indicates a constant fraction that is required to tell the difference between the weights.

An outcome of Weber's meticulous experiments was the grounded data he obtained to identify certain distinct capacities in the sense of touch. The foremost of these are location, pressure and temperature. Besides the touch organs, the voluntary muscles along with the nerves and their central organs provide a concept of the extent of the pressure or tension exerted on the body. When weights are lifted, there is an exertion of the muscles, but if weights are placed on the fingertips, the judgments arise from the sense of touch in the skin. The density of nerve fibres in any region increases acuity.

4.3 Touch receptors and neural pathways

The haptic system employs a variety of neurophysiological mechanisms to yield a percept. The physical system induces a deformation pattern on the skin and the moving fingers. Movements of the hand, arm and shoulder change the length of muscles and the angle of joints. These changes and deformations are picked up by sensory neurons known as mechanoreceptors. These are cells that are specialized to convert mechanical deformations into an electrochemical response. The response in turn evokes a nerve impulse. Mechanoreceptors possess receptive fields that are sensitive to changes that vary both in space and time. The signals from the receptors are conveyed to the spinal cord and the brain. The human body contains a variety of mechanoreceptors. Sherrington (1906) proposed a classification of receptors that is still in use. Stimuli at the surface of the body are conveyed by exteroceptors or cutaneous receptors. Deformations in muscles, tendons and joints are conveyed by proprioceptors. Mechanical deformations of internal surfaces like blood vessels and organs are conveyed by interoceptors. Cutaneous receptors are further classified by their anatomical structure. The cutaneous mechanoreceptors that are relevant for the detection of deformations with the fingertips are the meissner, merkel, ruffini and the pascinian corpuscles. In the upper layers of the skin lie the merkel and the meissner corpuscles, while the ruffini and the pascinian corpuscles lie in the deeper layers of the skin. The receptor organs in the upper layers of the skin are more receptive to low frequency vibration, while those in the deeper layers are more sensitive to high frequency vibration. The pascinian corpuscles and the merkel cells are rapidly adapting, whereas the meissner and ruffini corpuscles are slowly adapting. The merkel cells and the meissner corpuscles have small receptive fields, whereas the ruffini and the pascinian corpuscles have large receptive fields. Two main pathways carry cutaneous information to the somatosensory areas. They are (1) the spinothalamic pathway (information from pain, temperature and diffuse touch) and (2) the leminscal pathway that mainly carries information from tactile and kinesthetic receptors via the spinal cord and dorsal column and synapse in the lower medulla. Most fibres cross over to the contralateral side at the brain stem, ascend in the medial lemniscus to synapse in the thalamic relay nucleus and project from there to the somatosensory areas in the cortex. Touch threshold measures correlate highly with the density of distribution of touch receptors in the skin and the somatosensory areas to which they project. Acuity is crucial for active touch. Pressure is externally regulated and feeling disappears unless there is movement (Sherrick & Craig 1982). Movement is a necessary component of active touch.

4.4 Inter-sensory integration

Undoubtedly, no sense modality functions in isolation for long. The sense of touch must be understood in relation to other modalities by cross-modal comparisons, since the components of touch have not been specified. E. J. Gibson (1969) emphasized that successful crossmodal matching relies on the pickup of invariant information which is not tied to any specific sense modality as colour is to vision or sound is to audition, but is amodal and nonmodality specific. However, cross-modal matches between touch and vision are usually inferior to within modality visual matching because of an absence of correlation between what is seen and touched. Therefore the notion of what is amodal or invariant needs more clarification. J. J. Gibson (1979) proposed that we learn by orienting ourselves to the environment and derive meaning from stimulation or affordances from stimuli. The notion of affordances would imply that we attribute meaning to stimuli according to the context in which they occur and therefore, the environment or its ecology sustains perception and learning. This also suggests that the pickup of information is not confined to a single sense modality but relies on the integration of information from different senses. Therefore studies of cross-modality matching provide relevant data for knowing aspects of the tactile modality.

Theoretical notions of the emergence and development of the sensory systems were mainly two- the unitary and the separatist views. The unitary view held that the senses initially function as one and get differentiated over development, whereas the separatists believed that the senses provide different information initially. Information is later integrated over development as a consequence of perceptual learning and matching between modalities.

Studies of crossmodal matching have revealed that characteristics of information perceived in one modality, for example touch can be perceived by another modality, such as vision. That is one can feel the smoothness of an object just as it can be seen. The table looks smooth and also feels smooth. Experiments that demonstrate an equivalence between senses for the cross-modal perception of form have indicated errorless identification of forms from the visual modality to the tactile modality (Krauthamer 1968; Lobb 1965; Abravanel 1971), although intramodal visual matching has usually been found to be more accurate than intermodal touch -visual matching. Therefore in cross-modal matches between vision and touch, vision seems to produce more accurate responses.

Despite the quantitative differences in vision and touch in the perception and discrimination of forms among sighted individuals, cross-modal transfer of information has been found to occur very early. Bruner and Koslowski (1972) found that infants as young as 8 to 22 weeks made appropriate grasping motions to only one of two balls that was small and graspable and tended to ignore the larger ball. It seemed that information about size seemed to be coordinated with prehaptic activity. These results also suggest that cross-modal matching need not be mediated by language. Bryant, Jones, Claxton and Perkins (1972) showed that prelinguistic infants as young as 8 to 9 months reached for objects they had touched, but tended to ignore objects they had not touched, suggesting that the matching between the two modalities was an outcome of correlating objects that had been touched with those that had been seen. There is even more compelling evidence from studies with apes (Davenport & Rogers 1970) that they can select by active touch previously unfamiliar objects such as handles, clamps and springs that were presented visually. In a subsequent study, Davenport and Rogers (1971) reported that apes can even select by haptic touch objects that match photographs, indicating that cross-modal equivalence is possible in the absence of verbal mediation. Although young children might prefer to verbalize stimuli (Gaydos 1956) by giving names of people like Art, Bill etc, to card board forms they see and touch subsequently, we have found (Ittyerah & Broota 1983) that matching of two dimensional random forms is sensory specific both in the visual and haptic modalities, unless subjects are encouraged to verbalize the stimuli (Ittyerah 1984). Adult subjects were required to match four sided random figures (see Figure 4. 1) within (v–v and t–t) and between modalities (t–v and v–t). Intra modal matching was always faster and more accurate than inter-modal matching, indicating the processing of sensory specific information.

Figure 4.1. Stimuli: Twelve four-sided random forms
Reproduced with permission of publisher from Ittyerah, M., & Broota, K.D. Inter-and intra-modal processing of sensory-specific stimuli. Perceptual and Motor Skills 1983, 56, 507–517© Perceptual and Motor Skills 1983

Thus perceptual matching between modalities is evident and vision does not necessarily dominate touch. In fact, studies on the dynamics of the hand in early infancy (Hatwell 1987) have indicated that the hand is both an executive as well as a perceptual organ and that early in infancy perceptual discriminations are facilitated by touch than vision. It is only after the age of five or six months that vision tends to dominate and infant's discrimination abilities tend to be guided by vision.

4.5 Unity of the senses

The unity of the senses reveals itself in a variety of ways: in sensory processes when the psychological behaviour of different senses show similarities; in perceptual processes when different senses provide common information about the world with characteristics of sensory experiences resembling one another, and in cognitive processes when language mediates thought for example as in verbal metaphors that describe or suggest similarities among sensory phenomena such as 'the bloody sun at noon' (Marks 1978). Correlations between the senses are evident in observable behaviour. Similar correlations can also be found in the neurophysiology of sensory systems. The central nervous system could be activated by the stimulation of two or more senses. Nonspecific multisensory neurons can be found all through the brain, in the association areas of the cerebral cortex known to be responsible for complex cognitive functions. For example, Murata, Cramer and Bach-y-Rita (1965) discovered neurons in the visual cortex of cats that responded not only to light, but also to sound and to a pinprick on the skin.

Neurons that react to light, sound and touch have been located in subcortical sites as the cuneate nucleus (Atweh, Banna, Jabbur & To'mey 1974). As early as 1977, Rebillard, Carlier, Rebillard and Pujol (1977) reported that the primary auditory cortex in congenitally deaf cats is driven by visual stimuli. Conversely, studies on the microphthalmic mole rat (*Spalax ehrenbergi*) have shown that auditory stimulation can drive neurons in the primary visual cortex (Bronchti, Heil, Sadka, Hess & Scheich 2002). Cells in the primary visual cortex of visually deprived mice, rats, cats, or monkeys respond to somatosensory or auditory inputs, suggesting cross-modal reorganization (Toldi, Rojik & Fehér 1994). For example, neurons in visual cortical areas 19 in monkeys with early visual deprivation respond to somatic inputs such as manipulating the experimenter's hand in search for food (Hyvarinen, Hyvärinen & Linnankoski 1991).

It is now known that some form of intersensory matching is possible during early infancy before the complete development of the cortex. Studies of the ability to recognize the synchrony of speech and mouth movements (Campbell & Dodd 1987) as well as matching objects that had been previously touched by vision (Bryant et al. 1972) indicate that the senses converge very early in development. Neurophysiological evidence shows that there are multimodal and bimodal neurones that fire to stimuli in more than one modality, in several areas of the cortex (Duhamel, Colby & Goldber 1991) though the nature of their function is not yet clear. Neurophysiological substrates of intersensory processing are mainly the association areas of the cortex and the convergence of inputs in single neurones in the posterior parietal regions. The posterior parietal cortex is situated near the visual and somatosensory regions of the cerebral cortex and there is evidence to show convergence between vision and somatic sensations in that area (Robinson, Goldberg & Stanton 1978; Leinonen, Hyvarinen, Nyman & Linnankoski 1979).

The importance of the parietal cortex for the integration of movement and visual information has been stressed repeatedly (Palliard 1991; Stein 1991). Stein (1991) argued that the posterior parietal cortex brings together proprioceptive, auditory, vestibular and limb movement information as well as visual and motivational signals; this creates a network for transforming sensory signals into signals which are suitable for motor control. Berthoz (1991) stressed that the representation and control of movements depends on multimodal inputs. Visual, vestibular and somatosensory information come together in the parietal (parietal-insular) cortex, which is concerned with the representation of head movements in space (Akbarin et al. 1988).

Neuropsychological evidence indicates the possible convergence of brain regions corresponding to particular modalities. These studies mainly deal with animal or human subjects that have been temporarily or permanently deprived of a sense modality. The expectation is that if the subjects though sensorily deprived are able to perform like normal controls, then it is an indication of convergence of the remaining intact modalities that enable performance. Hyvarinen, Carlson and Hyvarien (1981) sutured the eyelids of monkeys soon after birth so that they had no shape perception, though some light perception was available. When the eyelids were opened after twelve months and they were tested for visually directed reaching, some firing was observed in the cells of the visual association cortex, though that was not the case with control animals that had not been visually deprived since birth. This indicates a form of compensation by the adjoining cortical areas. The loss of a particular sense leads to the invasion of the deprived cortical area by inputs originating from other modalities, illustrating the capacity of the cerebral cortex for plasticity and reorganization (Ptito & Desgent 2006; Pietrini et al. 2009; Merabet & Pascual-Leone 2010).

Studies in animals have shown that when the brain is deprived of visual input at an early age, tactile and other nonvisual information is re-routed to the visual cortex (e.g. Hyvarinen et al. 1991; Chabot et al. 2008). This claim is largely supported by results of functional brain imaging studies showing activation of visual cortex in early blind subjects during Braille reading (Sadato et al. 1996; Burton et al. 2002) and other forms of tactile stimulation (Burton, Sinclair & McLaren 2004; Ptito, Moesgaard, Gjedde & Kupers 2005). Effects of visual deprivation in early life revealed brain activity patterns in occipital cortex evoked by tactile stimulation are significantly stronger in early blind compared to late blind subjects (Cohen et al. 1999; Burton et al. 2002). A recent study reported overlap of auditory and tactile information in the occipital cortex of early blind subjects (Renier et al. 2010). What this evidence indicates is that sensory neurons cannot be localized just as the understanding of behaviour is dependent on the integration or convergence of sense data from different modalities.

Are visual representations mandatory for object recognition? Kupers, Pietrini, Ricciardi and Ptito (2010) used fMRI to measure brain responses in a group of blindfolded sighted subjects while they performed non-visual object recognition tasks. Tactile recognition of face-masks and man-made objects of daily use elicited distinct category-specific patterns of neural response in the extrastriate ventral temporal cortex, that were similar to those elicited by visual recognition (Haxby et al. 2001) of the same object categories (Pietrini et al. 2004). Furthermore, the neural response patterns elicited by tactile perception of bottles or shoes, correlated with those evoked by visual perception of the same object category, indicating that these neural response patterns are supramodal in nature; that is, that they are not merely restricted to visual perception (Pietrini et al. 2004, 2009). Congenitally blind individuals showed category specific neural response patterns in the ventral temporal extrastriate cortex during the same tactile object recognition task that were similar to those measured in sighted controls. These findings indicate that in sighted individuals visual and tactile object perception activate the dorsal part of the lateral occipital cortex (LOC proper; Amedi et al. 2001, 2002; James et al. 2002) by showing a cross-modal correlation of response patterns between the two sensory modalities. The findings in the congenitally blind subjects are important because they indicate, category-related representations in the extrastriate visual cortex does not require visual experience. Experience with objects acquired through other sensory modalities appears to be sufficient to support the development of these patterns. Thus, at least to some extent, the visual cortex does not require vision to develop its functional architecture that makes it possible to acquire spatial knowledge of the external world.

All these findings indicate the convergence of different modalities in transacting with the environment. Therefore the notion that vision is most important in acquiring knowledge is a fallacy. Undoubtedly, vision provides information, but the point of contention is that vision is not the sole source of information and just as other modalities are dependent on vision to provide complementary information, vision is just as well dependent on sense data from other sensory modalities. This is evident in quite a few studies that have shown interrelationships among touch, vision and movement. For most part movement seems to be guided by vision. Vision also seems to influence the postures of sitting, standing or reaching in infants all through development. For example, Lee and Aronson (1974) found that the external context can affect the maintenance of postures in young infants. Infants aged about a year or near about the time infants begin to stand upright independently were placed in a room that was so constructed that the walls could be moved as and when required. When the walls before the infants were gently tilted toward them, the infants promptly lost their balance and fell backwards in the direction of the tilt. In fact, the direction of the tilt determined the direction in which the infants would fall. This indicates a relation between the external reference provided by vision and the maintenance of posture. It is of relevance to know that visual control seemed to be more important for the maintenance of balance than an intact skeletal structure with adequate muscles to resist the force of gravity. The experiments of Lee and his associates (Lee & Aronson 1974; Lee et al. 1983) showed that vision has proprioceptive functions and play a role in the stability of posture and movement. This has been observed when infants were found to sway with a moving frame (Butterworth & Hicks 1977; Berthenthal & Bai 1989) though their gravitational information and physical support were not altered. Concomitant visual and vestibular information play a major role in controlling upright posture and equilibrium in locomotion among human adults (Assaiante & Amblard 1992). Balance in human locomotion depends upon stable head posture with regard to the body and on gaze relative to fixed landmarks in external space. Assaiante and Amblard (1992) showed that there is a stable adult head position during locomotion by the age of eight years. What these findings suggest is that information from more than one sense modality or correlation between sense modalities is a necessary process during perceptual learning and development.

Touch and movement are good examples of convergence of information between sensory modalities. In fact, without movement, the touched object cannot be clearly processed. My emphasis is on the relevance of different sources of information that complement each other to integrate processing in a largely unspecified modality like touch. Blind infants have been found to respond to touch cues rather than sound cues (Bigelow 1986) indicating that sound sources are less preferred than actual touch in alerting the presence of objects. This is further supported by Hatwell's (1987) findings, that early in infancy touch precedes vision and it is only after the ages of six months that infants begin to rely more often on vision and therefore utilize their tactile abilities to a lesser extent. Other instances in which the modalities have been found to converge is the usual synchrony between the sounds of words in speech and lip movements used in producing them (Dodd 1977). Even in noisy conditions, the advantage of seeing lip movements in producing words improved their being heard correctly. This has practical implications for the use of lip reading in the profoundly deaf (Dodd & Campbell 1987).

Therefore the modalities provide specialized but converging information. In the absence of some major source of information, informational redundancy is relevant. Millar (1986a) for example found that redundant information could improve tactual recognition of otherwise difficult patterns. Braille characters that are usually difficult to identify for beginners were made easier by increasing the number of dots or dot density with the shape or size of the pattern, and this difference was found to improve discrimination by both blind and blindfolded subjects. Redundant information is generally useful in conditions of uncertainty. When tactile cues are not sufficient, improving them with redundant information by repeating characteristics of the original stimulus in the form of size or shape has enabled recognition (Millar 1986a). Some modalities may be more correlated than others. But that does not mean that they provide the same information. Rather modalities converge or covary and errors or illusions occur when the normal covariance of different inputs is not compatible.

4.6 Studies with blind subjects

Evidence from studies with congenitally blind subjects has revealed a wealth of tactile and cross-modal information. Even though sighted blindfold subjects are treated as cohorts in specific studies, the sensory experiences of the two groups differ and are not comparable. The nature/nurture dichotomy has long been an argument in developmental science. The question of whether ability is innate or acquired has been often raised in studies with the blind. The famous Molyneux question, whether a man born blind and made to see, would be able to recognize by vision alone, objects and spatial dimensions which had previously known by touch was posed at John Locke (1975) who assumed that concepts derive from sensory experience. Locke's empiricist philosophical theory differs from earlier Platonist notions of innate (perfect) ideas, and with the eighteenth century rationalist argument by Kant (1781) that it would be logically impossible to perceive the world except in terms of a priori concepts of space, time and causality. Molyneux's question has raised speculation and suggested that there is an empirical test for the philosophical empiricist- rationalistic controversy. If spatial concepts derive from sensory experience, then the empiricist position can be considered to be true because the blind cannot have spatial concepts in the absence of vision since vision predominantly provides information about space, whereas if the blind can solve tests of spatial concepts, then it must be assumed that spatial concepts are innate and therefore sensory experience is not required.

Millar (1988) has cogently argued that the main reason that such a direct test of philosophical controversy is a mistaken one is that the innate/acquired dichotomy makes a number of empirically unjustified assumptions about the functioning and interplay of sensory modalities. The first fallacy is the assumption that vision is the only sense modality that provides information of three dimensions in space. The second assumption is that different sense modalities convey different information that cannot be integrated. Thirdly, the innate/acquired dichotomy is usually also taken to imply that the sense modalities are separate and require some mediation (e.g. Connolly & Jones 1970) or a translation mechanism. Or, that the sense modalities are unitary and provide the same information and the only other alternative is that ideas are innate. Fourthly, the dichotomy assumes that the relation between perceptual and conceptual cognitive skills must be hierarchical, that is, either data driven from sensations to concepts or 'top down' from concept to perception. However none of these assumptions are justified (Millar 1988) and vision is not the only modality that provides spatial information because spatial information can be derived from hearing, touch and movement. Besides, even if there is modality specific information, there is an interaction between the modalities as they process information from the environment and sensory information is normally convergent and complementary and can be a part of information processing (Millar 1981a).

Speculations as to whether the blind have spatial ability suggest that knowledge of space mainly exists only for the visually informed (Revesz 1950; Warren & Pick 1970; Warren 1977). According to Revesz, vision is the only spatial modality and provides information about external space, whereas, haptics (touch and movement) is the only source of spatial information for the blind and provides information about personal space. However, the potential for learning about space in the absence of vision cannot be ignored, and if there is a potential, it is natural that the ability can develop. Ittyerah and Samarapungavan (1989) demonstrated in a series of Piagetian tasks that congenitally blind children like their sighted cohorts were able to perform tests of conservation, seriation, rotation and transitivity over different ages of development. Though the blind found the conservation of substance task difficult as compared to the other tasks, they were better than the sighted blindfolded on the seriation task indicating their familiarity with processes of seriation or successive processing of information in the absence of vision. The congenitally blind also performed on the mental rotation tasks and showed an effect of development in the performance of the older children. Therefore development of spatial notions is possible in the total absence of vision.

Millar (1976) demonstrated that though the blind do not have access to all the spatial information experienced by sighted people, the blind rely on different heuristics and decoding strategies in attaining spatial information. For example, adopting a body centered frame of reference so that the stimulus is aligned to the mid line of the body from which the distance, direction and location of a target within personal space can be assessed by the moving arm and fingers is more prone to error than geometric heuristics in mental rotation and re-location problems. However errors do not depend on the strategies that are used, because such body centred spatial strategies can be updated as and when required. Further, coding by movement and coding relative to a body centred frame has been experimentally distinguished (Millar 1981b). The notion that an egocentric frame of reference indicates a lower level of development than an allocentric frame of reference is not true. Findings that even nine month old infants code object locations in relation to external cues in experimentally appropriate conditions (Bremner 1978) and the fact that the upright posture of sighted infants is guided by the external visual frame than by gravitational proprioceptive information (Lee & Aronson 1974) indicates that the assumption that spatial development consists of a change from body centred reference frames to external Euclidean coordinate frames is false. In Lee and Aronson's study, standing infants fell to the floor when they were confronted with a tilted screen showing a loss of balance when their external frame of reference changed from an upright wall to a tilted screen. The fact that the blind can and do solve spatial problems (Leonard & Newman 1967; Millar 1976) and that the sighted do make errors in spatial tests, indicate that vision is neither necessary nor sufficient for spatial coding. In Leonard and Newman's (1967) study blind children were able to arrive at a desired location by taking a different path from the one they had taken initially. This indicated that the blind children were able to represent space as they moved from point to point and utilize that information to reach the target place.

This does not mean that spatial concepts are innate (Landau, Gleitman & Spelke 1981; Landau et al. 1984) as was suggested by the performance of Kelli, a two and a half year old congenitally blind child, trained to walk between four landmarks in a diamond shaped configuration. The possibility that Kelli could have relied on other sources of information such as light perception with which she could distinguish light from shade as she moved, or feedback from her own footsteps, since she was able to correct her movements if she started in the wrong direction, cannot be ignored (Millar 1988). Further, Kelli's paths to the target were curved and this is not compatible with Euclidean principles (Millar 1988). If Kelli was able to perform the task, it was because she relied on information from sources other than an innate ability to represent Euclidean space. On the other hand, the development of space can occur by adopting a frame of reference from which spatial distances can be compared, either relative to the center of the body or an external object. Millar (1986b) found that congenitally totally blind children of about ten years of age used a similar two dimensional scheme for drawing a human figure as their sighted peers. Younger children used scribbles, though neither groups of blind children had drawn before. This does not mean that their drawing schemes were innate, but rather show an attempt on the part of the blind children to translate from 3-D to 2-D space by using shapes to represent body parts. The older blind placed parts of the figure randomly on the page. It showed that at the very first time they could not think as to how they should represent a figure relative to 3 D space. This does not mean that they lacked the potential to solve the problem. Sighted children adopted a frame of reference by placing the figure parallel to the side of the page so that it can be used to indicate the vertical upright in space. The blind can also adopt similar solutions. Millar (1986b) suggested 2D spatial displays could be used to teach 3 D relations to blind children.

4.7 Haptic cognition

The active movement of the hand about the object involves the conjoint information of surface properties such as texture and temperature from the skin and touch receptors of the hand, and the application of pressure or force to the object. Touch modality conveys information about a number of different, although often overlapping skin sensations. Inputs can arise from vibration or temporally spaced pulses, which give impressions of hardness or softness, from patterns of roughness or smoothness and dry or wet surfaces. These impressions can be used to identify objects that give spatial information about extended surfaces and the spatial relations between them. Therefore touch is not a slow impoverished modality as compared to vision (Millar 1994) and the hand is considered to be a unitary sense organ like the eye.

The processes of vision and touch must be distinguished. One such distinction is between identifying objects and specifying their spatial orientation, features and locations. For vision there is evidence that processes differ for tasks that ask about the nature and function of the object as compared to tasks that require its location or shape (Weiskrantz 1986). Descriptions of touch refer to tasks that involve identifying objects by their prominent features so that they may be recognized by their categories and function (Lederman & Klatzky 1987). Although knowing the shape of an object can help to determine its function, recognizing a rotated or a displaced object requires knowing more than its shape. Shape perception depends on spatial organization of the stimulus wherein the felt features have to be located by reference to each other or to some internal or external frame. Rotated objects have to be identified in relation to a coordinate frame. Therefore identifying objects requires the ability to relate the object to a particular frame of reference in space and not only the movement of the hand according to stimulus characteristics of shape or size.

Millar (1994) observed that in touch, the complementary information from other sources or modalities which is needed to identify a shape differs with the size and type of object. This has important consequences for a theoretical description of haptic coding. To recognize objects by touch across differences in size, depth and composition requires convergence of different complementary information. Paillard (1971) suggested that coding locations differ from shape coding in touch as well. Touch is a complex sensory modality that involves movement of the limbs besides other sensations such as of temperature or pressure. The motor cortex represents limb movements from the head downwards with larger areas devoted to the lips and fingers in primates and humans. The interconnections between the motor cortex and the cerebellum are crucial in voluntary movement, movement learning and motor control and are also connected with areas that are involved in visuospatial processing. Movement in space is also related to the way in which the tactile system adapts itself to the environment. Awareness of space relates to two main reference systems, one the self- referent system whereby the individual usually relates his body midline with the object or location in external space, and the other an allocentric reference of space based on knowledge of external space. It has been found that the frontal cortex, more specifically the pre frontal areas are interconnected with the parietal lobes as well as with the subcortical areas and these are also involved in the representation of space. What this evidence indicates is that the tactile system cannot be studied in isolation because any act of movement has an effect on the tactile system (Millar 1994).

Crossmodal effects or the matching of information between two different modalities such as visual inspection of the stimulus and tactile recognition of the visual stimulus, or the tactile inspection of the stimulus and visual recognition, are not symmetrical and produce different effects (Connolly & Jones 1970). Inputs from different sources can produce amalgams that differ from either. For example, if different syllables are presented simultaneously to hearing and vision, they may be heard as a syllable that differs from both (Campbell 1987), indicating that both auditory and visual information contribute to perception. Crossmodal perception suggests that information from one sense modality (touch) can be transferred to another (vision) for recognition of the stimulus initially inspected by the previous modality (touch). Since this process involves visual effects, tactile perception can be influenced or contaminated by visual effects. In order to understand tactile processes, one must test for information in the absence of vision, such as in congenitally blind conditions. Therefore haptic cognition hence forth in this manuscript will be considered as a process in its own right, unaided by vision.

According to Millar (1994), perception by touch depends on complementary information from tactual acuity, active movement and spatial cues that are not the same for all shapes. Davidson (1972) and Berla and Butterfield (1977) showed that shape perception depends on systematic exploration. Haptic perception of shape properties depends on the type of exploratory activity that is deployed (Appelle et al. 1980; Millar 1987b). The range and type of information needed for shape recognition differ among different sizes and types of object. The differences involve the availability of reference cues. The description of processes in tactual recognition differs for different types and sizes of shape, in a way in which it is not true of visual shapes. For example, description of haptically perceivied three dimensional objects is different from raised dot patterns and flat objects that are felt passively from being placed on the skin. In touch, as in vision, shape perception depends on the nature and kind of complementary information. But in touch that complementary information differs for different types of tactual patterns with the size of the object, continuity of stimulation, the means of exploration, and with the task demands and prior knowledge (Millar 1994).

Tactual recognition of objects may be found to be inferior because tactual inputs are not organized spatially in the manner visual information is presumed to be (Millar 1986a). For example, Wolfe (1966) observed a blind twin feel the length and breadth of a wooden spoon with much accuracy as compared to its sighted twin. This indicated that the nature of haptic processing is characteristically serial (Ittyerah & Samarapungavan 1989) and when sighted subjects are required to seriate haptically the global nature of visual processing interferes with haptic seriation.

Studies have shown that blind subjects without access to visual codes can match tactual features of verbal inputs (Millar 1977b). The effects of distractors on tactual memory indicate that tactual memory is affected by attentional demands (Gilson & Baddeley 1969; Millar 1974) as visual memory (Posner, Boies, Eichelman & Taylor 1969). Millar (1975b) found that recall decrements on tactually similar serial items were shown mainly by subjects testable on smaller set sizes, while subjects testable on larger set sizes showed decrements mainly when serial items were phonologically similar, suggesting that processes underlying tactual memory for serial items differed from verbal memory. Millar (1977b) observed that tactually presented verbal material can be represented in memory. Besides naming, tactual features of the verbal input can also be represented in memory for brief periods. For example, Watkins and Watkins (1974) demonstrated the tactile suffix effect. The fingers of the subject were touched in a sequence after which this sequence was interrupted by the experimenter, by stroking the fourth finger for example; it was observed that subjects were unable to recall the sequence in which the fingers were touched before the stroke. Therefore, the stroke served as a mask that disrupted the 'recency effect' in touch showing tactile information can be retained for brief periods. Findings (Millar 1977b) showed that tactual features of verbal stimuli are used in memory and memory is not confined to naming alone. Matching successive letters in the same physical format was faster than matching the same letters when the standard was in a different format. Tactual characteristics of letters in memory affect Braille letter recognition. In a subsequent study Millar (1978) tested the hypothesis that processes underlying memory for tactual shape differ from those underlying memory for verbally coded material by comparing blind children's recall of tactual nonsense and letter shapes under conditions of grouped and ungrouped presentation. Results showed that grouping significantly impaired the recall of nonsense shapes, while letter recall was better and improved significantly with grouping by subjects with larger recall spans whose pre test letter naming was relatively fast. For subjects with smaller letter spans whose pretest letter naming was slow, letter recall did not differ from nonsense shape recall. The processes underlying memory for the recall of tactual inputs that are difficult to name differ from processes determining recall of easily named material. The feel of shapes is less remembered as compared to names of shapes. The processes underlying short-term retention vary with the form of coding. Processes underlying the recall of tactual shapes were subject to decay rather than to attentional demands, indicating that the form of coding may be one factor in determining the nature of processing and recall efficiency (Millar 1978).

The haptic perception of an object involves touch and movement. Movements are produced not only actually as the movement of the hand about an object, but movements can also be mentally represented. Therefore there could be movement imagery, just as one can mentally produce the visual image of an object. Findings that congenitally blind children are better at representing spatial configurations by drawing movements than at recognizing the same figures indicated that movement imagery can be learned and is based on output plans (Millar 1985b, 1986b, 1991). These findings showed that nonverbal movements can be attended to repeated and reproduced and can mentally represent nonverbal materials. We (Millar & Ittyerah 1991) have shown modality specific movement imagery in the absence of visuo spatial information. The blind like the sighted blindfolded children were able to reproduce an aiming movement both after actual practice and mental rehearsal of the linear extent. Prior to recall subjects either had to perform or to imagine irrelevant (smaller/larger same) movements. Irrelevant movements whether actually performed or imaged produced significant bias for both the blind and sighted children, although bias was smaller in imagery condition. Performance improved with rehearsal indicating that visuo spatial imagery was not a necessary condition for the cognitive skill. Further, mental practice effects may be mediated by more than one factor. It is unlikely that mediation was verbal, since articulatory suppression (repeating an irrelevant syllable during the delay periods either aloud or subvocally, mouthing it) had no effect on any group, suggesting that subjects did not translate the target movement into some verbal description that could have been repeated subvocally. Therefore movement imagery has been demonstrated in the total absence of vision. Thus mental practice effects were evident in the total absence of vision. These findings should stimulate research that would enlighten the role of mental processes in haptic cognition.

4.8 Multimodal spatial interactions

Multimodal spatial interactions are common in daily activity as the individual integrates sounds with sights or tactile sensations in the mouth with smell and taste. Just as sounds are heard and colours are seen, touch informs the individual about surface texture, pressure, temperature and spatial characteristics of objects. There is evidence that as a tool is held in the hand and used to reach an object on a table for example (Maravita, Spence, Kennett & Driver 2002) the tool extends the space for tactile interaction with objects. Since the tool can weild several postures that the hand may adopt to contact the object, the spatial mapping between tactile stimulation at the hand and any related visual information at the end of the tool alters as a tool is used. Maravita et al. found that prolonged active use of tools for several minutes in human adults can modify visuo tactile spatial integration and create a progressive functional change in body schema, so that with experience in using the tools they become treated as extentions of the effectors wielding them and show strong integration of visual information at the far end of the tool with tactile information at the hand wielding the tool. Iriki, Tanaka and Iwamura (1996) demonstrated that when monkeys were trained to use tools to reach objects they could see in far space, the visual receptive field of such multimodal cells could be enlarged to include the whole length of the tool. As a consequence bimodal neurons that respond to stimuli near the hand begin to respond to visual stimuli at the far end of the tool. When Iriki, Tanaka, Obayashi and Iwamura (2001) introduced different mappings between hand location and the location at the tool end, corresponding remaps were found in the bimodal cells after considerable practice.

Multisensory interactions exist between vision, touch and proprioception (Rizzolatti, Matelli & Pavesi 1983) in relation to the perception of limb position and the peripersonal space surrounding the limb. Single cell studies in primates have revealed populations of multi- sensory cells in several brain areas that have spatial properties (Graziano & Gross 1995; Graziano & Gross 1998; Hyvarinen & Poranen 1974). Two human fMri studies (Macaluso, Frith & Driver 2002; Lloyd, Shore, Spence & Calvert 2003) showed that depending on the posture of hand placement or eye direction, a given vibrotactile stimulus can produce different brain activations as a function of the sight of the hand or visual stimulation near it. Thus multi- sensory interactions between vision and touch have been demonstrated in single cells with animals, healthy humans and brain damaged patients (Maravita, Spence & Driver 2003). In all instances, the multisensory interactions between tactile stimulation on the hand and visual stimulation of objects in space depend on the proximity of the visual stimuli to the relevant body part in external space. These crossmodal effects respond to changes in posture and remap according to the spatial combination of receptors in different modalities that need to be stimulated to produce alternate postures.

4.9 Three dimensional shapes

The world about us consists of varying sizes and shapes of objects, some too small to be detected by the human hand and others too large for haptic exploration. The duplex theory of touch (Katz 1989), states that very fine textures are processed temporally whereas more coarse surface structures are felt spatially. The human hand can actively explore spatial scales of object sizes ranging from a few microns to several meters. Johansson and LaMotte (1983) for example found that sharp edges, only one thousandth of a millimeter high, could be felt with the fingertip. Perception of such microscopic objects reveals the powerful sensitivity of the tactile system and knowledge of the characteristics of sensitivity might help to understand how perception works.

Touching objects involves feeling ridges or protrusions, and also smooth undulations such as shapes of vases or sculptured faces. In order to test for the sensitivity of the tactile system for minimal differences between two undulating surfaces or the just noticeable difference between two surfaces, Nefs, Kappers and Koenderink (2001) measured discrimination thresholds for sinusoidal gratings using active dynamic touch. At first they measured the thresholds for amplitude discrimination as a function of amplitude and spatial period. Amplitude differences between gratings as small as 2 micrometers could be detected by the fingers. Weber fractions or the just noticeable differences for amplitude discrimination were constant over a range of amplitudes, but were influenced by the spatial period of the grating, in that discrimination improved with increasing spatial period. In a second study Nefs et al. (2002) found that amplitude has no effect on the Weber fractions for spatial period discrimination, though the spatial period had an effect on the Weber fractions, in that they decreased with larger spatial periods.

The psychophysical method of relating a perceived sensation such as heaviness or brightness to the actual physical value of the weight of the object or the intensity of brightness has yielded promising results in detecting thresholds or psychological indexes of values for discriminating between stimuli. For example, the experiments of Davidson (1972) were primarily concerned with the scanning strategies of blind and sighted subjects. He used 200 mm curved strips and varied the curvature of the strips in order that the participants could judge whether a strip was concave, flat or convex. He found that the discrimination thresholds of the blind were lower than that of the sighted, though the effect disappeared when they were restricted to the same scanning strategy. Gordon and Morrison (1982) used convex stimuli with scanning lengths of 20, 30 and 40mm. They measured haptic detection and discrimination thresholds and found that the thresholds were determined by the slope of the curvature of the stimuli, over the curved surface relative to a flat surface. Pont, Kappers and Koenderink (1999) studied curvature comparison with a range of cylindrically curved strips. The contact length of the strips was varied from 50, 100, 150 to 200 mm. Discrimination thresholds were found to increase linearly with increasing contact length. Pont et al. concluded just as Gordon and Morrison (1982) that the effective stimulus for detecting curvature differences between stimuli is the slope of the stimulus.

Stefan Louw (2002) in his Ph.D dissertation described a series of experiments in which he measured detection and discrimination thresholds with varying spatial width and amplitude. The thresholds were measured for a range of spatial scales such as a Gaussian profile, a Mexican hat or a sine wave grating. The width of the Gaussian profiles referred to as the spatial period varied between 150 micrometers to 240 mm. The results indicated that detection thresholds depend on the spatial width of the Gaussian profile He found a power law relation between detection thresholds and spatial width with a power factor of roughly 1.3 for both convex as well as concave surfaces. This power relation could be generalized to other stimuli that may differ in height and slope.

The findings of Stefan Louw suggest that the haptic system is sensitive to the maximum slope of the stimulus. Furthermore, when the amplitude and the width of the Gaussian profiles were varied simultaneously, discriminations were better for sharp (extensive amplitude and little width) from smooth (little amplitude and extensive width) Gaussian profiles, than for small (little amplitude and width) from large (extensive amplitude and width) profiles. These studies of detection and discrimination indicate the fine range of stimuli that can be perceived by touch.

To be able to detect and discriminate between various surfaces is a test of the sensitivity of the modality, but to be able to tell a vase from a face involves haptic cognition. This is an active process that combines the perceived features of the stimulus with knowledge of its category and function. Several studies have compared visual perception, tactile (haptic) perception, and visual-haptic perception of stimuli (e.g. Easton, Greene & Srinivas 1997; Millar 1981; Abravanel 1971; Lobb 1965; Rudel & Teuber 1964). Often, performance in tasks involving unimodal visual perception exceeds performance in both unimodal haptic and cross-modal tasks. But not always: Norman, Norman, Clayton, Lianekhammy, and Zielke (2004), for example, found that unimodal haptic comparisons of natural three-dimensional shapes could be as good as visual-haptic and haptic-visual comparisons. Norman et al. inferred that vision and touch have functionally overlapping, though not equivalent, representations of 3-D space.

A large body of literature concerned with the haptic and visual memory systems has provided two broad accounts. The first is that object memory is multisensory and object representations are shared across modalities if the same object properties are encoded. The second is that objects are stored as modality specific representations that require a recoding from one memory system to another. Such recoding is costly, in time and errors, to cross-modal performance relative to intra-modal performance.

Many recent findings support the idea that visual and haptic memory for objects is shared. For example neuroimaging evidence suggests that the cortical area involved in visual object recognition is also involved in haptic object recognition (Grill-Spector, Kourtzi & Kanwisheri 2001). Amedi, Malach, Hendler, Peled and Zohary (2001) reported that activation in the lateral occipital complex (LOC) is not specific to the visual modality. The LOC is also activated during haptic object recognition, though not by auditory information about objects (Amedi, Jacobson, Hendler, Malach & Zohary 2002). These studies suggest that the LOC is involved in recovering the geometrical shapes of objects (James, Humphery, Gati, Servos, Menon & Goodale 2002), and behavioural evidence suggests that object representations are shared across modalities (Easton, Greene & Srinivas 1997; Newell 2004).

Other studies indicate, however, that tactile memory is not equivalent to visual memory and that information is modality specific. For example, in tests of tactile memory, when a delay intervenes between the presentation of the test stimulus and the test of memory, tactile memory is adversely affected by verbal tasks (counting backwards) interpolated during a 15 sec delay (Gilson & Baddeley 1969) and by arithmetic tasks after a 5 sec delay (Sullivan & Turvey 1972).

Other studies report that tactile memory can be sustained for 15 sec (Kiphart, Hughes, Simmons & Cross 1992) and is vulnerable to articulatory suppression (Mahrer & Miles 2002). However, performance can also be affected by the task demands or stimulus complexity and the degree of familiarity with the objects being explored (Millar 1981). For example Norman, Norman, Clayton, Lianekhammy and Zielke (2004) observed that accuracy of tactile-visual matching differed across different stimuli, though tactile performance improved with exploration time. Norman et al. concluded that observers can match objects known only through touch with other objects known only through vision. Wood, O'Modhrain and Newell (2004) suggested that tactile-visual object recognition may rely on modality specific representations. Although the effect of delay between cross-modal presentations was the same whether the initial object was coded visually or haptically, recall was better after a 0 sec than a 30 sec delay, indicating delay-induced decay in memory.

In view of the different interpretations, Ittyerah and Marks (2007) tested for inter-modal and intra-modal processes involved in memory for concave curved surfaces that were first presented (perceived) haptically or visually and then later compared to surfaces presented either haptically/visually (unimodal comparison) or visually/haptically (cross-modal comparison) (see Figure 4.2).

Furthermore, by varying the characteristics of the subjects' activity during a 30-second retention interval (dual task design), we sought to illuminate the underlying mnemonic representation mediating the discrimination.

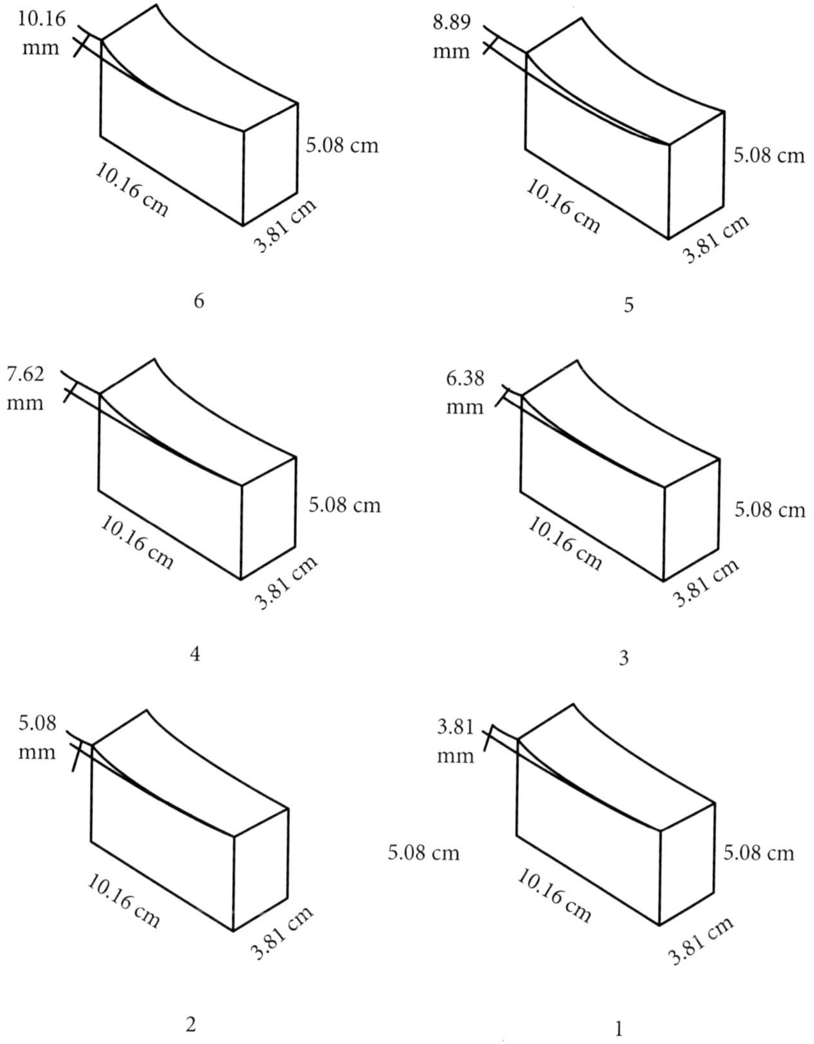

Figure 4.2. Stimuli

Reproduced with permission of publisher from Ittyerah, M. & Marks, L.E (2008) Intra-modal and and cross-modal discrimination of curvature: Haptic touch versus vision. Current Psychology Letters, 24, 1–11. Copyright © Current Psychology Letters, 2008

Dual task paradigms have been used to test for the demands of attention on the primary task (Brown 1958; Peterson & Peterson 1959). The paradigm tests the nature of coding in short term memory by interpolating a secondary task during the delay interval between presentation and recall. Dual task paradigms that test visuo-spatial coding often require movement outputs. For example, tracing, pointing or other gestures are used as secondary tasks. Imagining the task during a delay interval instead of actually performing the task produces similar effects. Millar and Ittyerah (1991) for example, showed modality-specific motor memory in blind conditions not only in actual performance but also in imagined conditions, which excluded any influence of visual knowledge. Thus, it is possible to use movement imagery in the recall of guided movements, as imagined movements biased recall as much as actual movements did. Further, articulatory suppression had no effect on performance, implying little or no role for translation of extent of movement into a verbal format. However, the nature of this modality-specific motor memory is still not well understood. The question arises, for example, whether motor memory is represented spatially in terms of spatial extent or extent of movement. Studies of movement have shown that extent of movement is encoded differently from spatial location in short-term memory (Laabs 1973; Laabs & Simmons 1981). Short-term memory could rely, therefore, on kinesthesis even when inputs are not coded spatially, although memory for extent of movement is not very accurate.

Are differences between haptic and visual perception of curvature evident in memory? We (Ittyerah & Marks 2007) expected that haptic, visual, and cross-modal memory will differ in their sensitivity to the activities in which subjects engage during the interference period. During a delay period of 30 seconds between the presentation of a test stimulus and its recall we introduced any one of the four following activities, such as counting aloud, rehearsing the test stimulus visually or haptically, spacing paper clips in equal distances or moving books from one hand to the other, as well as an unfilled delay that served as a control. We expected that tasks requiring spatial processing may have deleterious effects on performance in both modalities, whereas tasks requiring movement should exert greater effects on haptic memory. The findings indicated that performance was not only better in haptic than visual conditions, but also that the intervening activities exerted greater effects with haptic than visual presentations (see Figures 4.3 and 4.4).

Prior evidence has shown that vision improves shape matching and seems to dominate over touch and haptic inputs that involve touch and movement (Held 1963, 1965; Rock & Victor 1964). It was also generally agreed that vision is most important in spatial tasks (Sendon 1932, Attneave & Benson 1969). Furthermore two studies suggested that noninformative vision can improve perception by touch (Kennett, Taylor-Clarke & Haggard 2001; Newport, Rabb & Jackson 2002). These findings are not consistent, however, with the evidence that people who are totally blind from birth can be equally or more proficient than the sighted on spatial tasks (Hollins 1968; Millar 1994). This evidence implies that proficiency varies with the spatial information that is available from other sources. For example, in a spatial location task of six landmarks, Millar and Al-Attar (2005) found that vision affects haptic processing only if vision adds task relevant information. In experimental conditions of touch with diffuse light perception that excluded spatial cues and touch without vision, there were no differences in accuracy of performance. Millar and Al-Attar concluded that the differences between performance with spatially relevant and spatially irrelevant visual information provide new evidence against the hypothesis that vision affects haptic processing even if it does not add task-relevant information. Therefore, the relatively better performance of tactile compared to visual judgments (Ittyerah & Marks 2007) may reflect differences in the relevance of the delay tasks used to probe the tactile and visual systems.

It is conceivable that the differences in mnemonic representations for haptic and visual curvature are related to differences in haptic and visual perceptual processing. Haptic processing depends on movements of the hand or finger over the stimulus and therefore is subject to spatial constraints on motion, whereas visual processing is often global (Navon 1977) and the sensory detection and discrimination are limited by the physical properties of the stimuli themselves, such as photon counting at low intensities (De Vries 1943). In linear movement

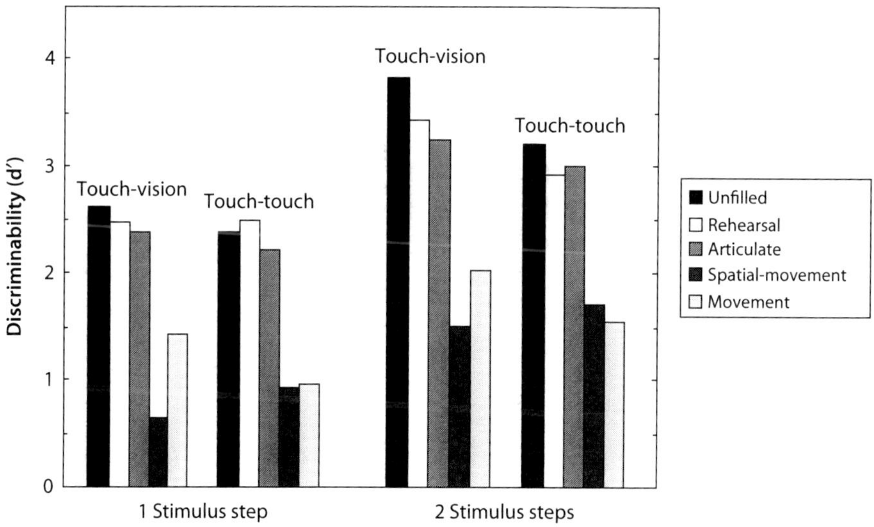

Figure 4.3 Discriminability (d′) as a function of the activity intervening between the first stimulus (always tactile) and the second stimulus (visual or tactile) for small differences (1 step) and larger differences (2 steps) in curvature

Reproduced with permission of publisher from Ittyerah, M and Marks, L.E (2007) Perception and Memory in Curvature stimuli. Haptic Touch versus Vision. British Journal of Psychology, 98, 589–610. Copyright © John Wiley, 2007

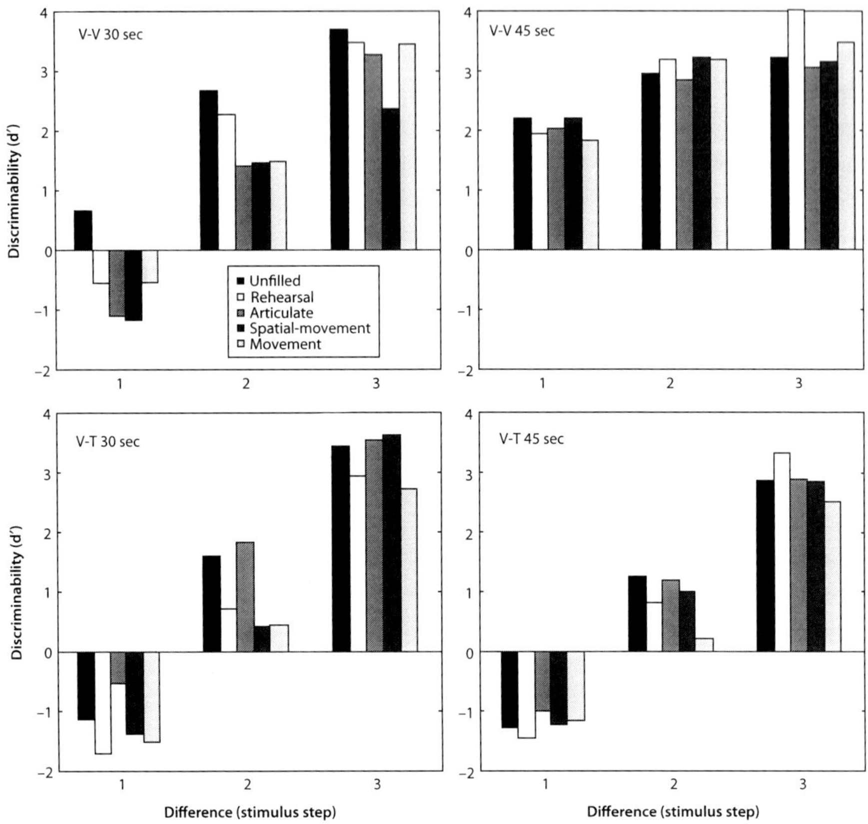

Figure 4.4 Discriminability (d′) as a function of the activity intervening between the first stimulus (always visual) and the second stimulus (visual or tactile) for small differences (1 step) and larger differences (2 and 3 steps) in curvature

Reproduced with permission of publisher from Ittyerah, M and Marks, L.E (2007) Perception and Memory in Curvature stimuli. Haptic Touch versus Vision. British Journal of Psychology, 98, 589–610.

tasks the demands on memory involve the starting and ending locations of the movement (Laabs & Simmons 1981; Millar 1994). For haptically felt curvature the demands on memory seem to be confined to the slope differences over the far ends of the stimulus (Gordon & Morison 1982; Pont et al. 1997; Pont et al. 1998) for both static and dynamic touch of curved strips (Pont et al. 1999). The findings of Ittyerah and Marks (2007) suggest, in turn, that memory for haptic and perhaps for visual representations of curvature, perhaps representations of slope, may be particularly disrupted by tasks that involve spatial processing and movement – as assessed with the dual task paradigm used to test for the demands of attention on the primary task (Brown 1958; Peterson & Peterson 1959). Filling the delay interval with a spatial or a movement task in the haptic and visual modalities produced modality-specific interference (e.g. Logie 1986). These results, though specific to the present tasks, may be generalized as being important characteristics in the perception of the curvature of objects. Besides, memory for information from touch and movement has been demonstrated by showing effects of coding texture for unfamiliar shapes and kinesthetic coding for unfamiliar movements (Millar 1999). This indicates distinct effects for tactile memory. Coding can also involve the mental rehearsal of movements (Millar & Ittyerah 1991) showing modality specific aspects of the input information. However the informational conditions in which such coding depends differ from those in which visual cues are present. Nevertheless, haptic representations are not recoded into visual coordinates, since impaired memory for recognition in one modality is generally dissociated from performance in the other modality (Farah 1990; Reed, Caselli & Farah 1996).

According to Norman et al. (2004), touch and vision are similar in that they are both sensitive to an object's global or overall shape (Norman, Dawson & Raines 2000; Lakatos & Marks 1999), though the human ability to perceive differences in local metric structure between objects such as depth and curvature is not accurate (Norman et al. 2004). Therefore there are important similarities between vision and haptics. This may suggest that the visual and haptic systems may be interconnected (e.g. Amedi et al. 2001). If touching an object should trigger some visual representation of it, then seeing an object might have a similar effect on the haptic system. However, the haptic recall of curvature was better with delay than was visual recall indicating different effects between modalities with delay. The results of the study indicate that visual and haptically obtained information about curvature is likely retained in both spatial and movement formats. This does not necessarily indicate that the representations are shared between modalities or that they are recoded from one modality to another, but could indicate that certain shape characteristics are perceived similarly within and between modalities. This outcome suggests possible differences between vision and touch in processes of encoding or representing stimulus curvature. It was of interest, therefore, to compare visual, haptic, and visual-haptic discrimination of these same stimuli when the two stimuli within each pair were presented simultaneously.

Ittyerah and Marks (2008) instructed subjects to explore two curvature stimuli with the forefingers of the two hands by simultaneously moving the fingers, in the same direction, from one end of the stimuli to the other in one continuous motion. A single continuous movement of the finger was considered to specifically focus on the curvature being felt, whereas free exploration of the stimulus with the entire hand might have distracted the participant by the irrelevant dimensions of the stimulus. There was no fixed time limit for a single continuous movement, although this was considered equivalent to a one second visual exposure of the same stimulus. The fingertip was positioned at the centre of the surface of the curvature and not at the edges, so that the finger could easily move from one end of the curved surface to the other. There were no restrictions in the movement of the elbow while the finger felt each surface. On each trial, one stimulus was placed on the left and the other on the right (separated by about 10 cm), the left-right positions for each possible pair counterbalanced across repetitions for each participant. The stimuli were the same as presented in Figure 4.2.

Pairs of stimuli were presented simultaneously at eye level. Again, on each trial, one stimulus was placed on the left and the other on the right (separated by about 10 cm), the left-right positions for each possible pair counterbalanced across repetitions for each participant. The visual stimuli were exposed for one second. The visual stimulus was placed above the haptic stimulus, and the participant examined the two simultaneously. On half of the trials, both the visual and tactile stimuli extended leftward from the midpoints of the surfaces upon which they rested. The participant simultaneously felt the haptic stimulus with the left hand in a single continuous motion and saw the visual stimulus. On the other half of the trials, the stimuli extended to the right of the midpoints of the surfaces, and the

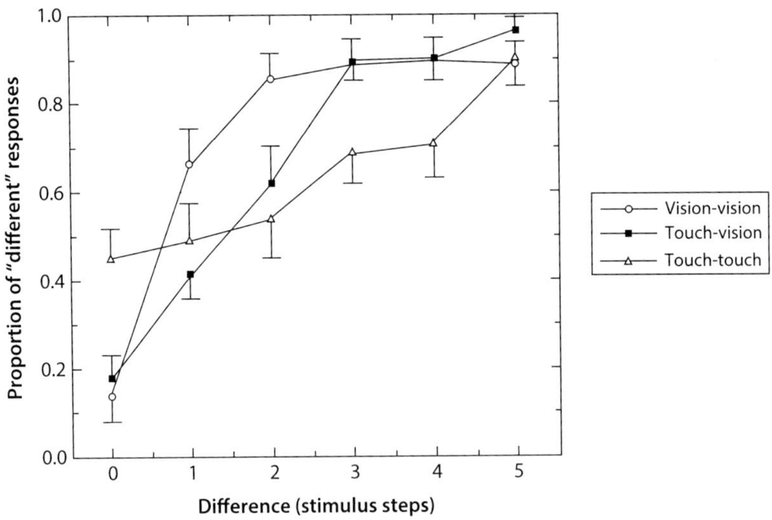

Figure 4.5. Proportion (and standard error) of 'different' judgments as a function of the size of the stimulus difference (in stimulus steps), for the visual-visual, tactile-visual, and tactile-tactile comparisons of curvature

Reproduced with permission by publisher from Ittyerah, M. & Marks, L.E. (2008) Intra-modal and cross-modal discrimination of curvature: Haptic touch versus vision. Current Letters in Psychology, 24, 1–11. Copyright © Current Psychology Letters, 2008

participant felt the tactile stimulus with the right hand. The averages and standard errors of the response proportions given to each pair of stimuli are depicted in Figure 4.5. Inspection of the figure reveals two general trends. (1) Performance tended to be best with visual presentation (V–V), worst with haptic presentation (T–T), and intermediate with cross-modal presentation (T–V), and (2) on those trials in which the stimuli were different, performance improved as the physical difference between the stimuli increased.

The second approach combined judgments given on same trials with judgments given on the various different trials into values of d prime. The value of d prime, from signal detection theory (Green & Swets 1966), provides a measure of discriminability of each pair of different stimuli, taking into account both the ability to identify stimuli of different pairs as different and the propensity to (mis)identify the stimuli of same pairs as different. Measures of d' thus take into account variations in criterion for responding 'different': The more lax the criterion, the more likely a subject will respond 'different' to a different pair, but also the more likely to respond 'different' to a same pair. The proportions of 'different' responses given to both same pairs and different pairs were then converted to d' (Macmillan & Creelman 1991). The values of d' are plotted in Figure 4.6.

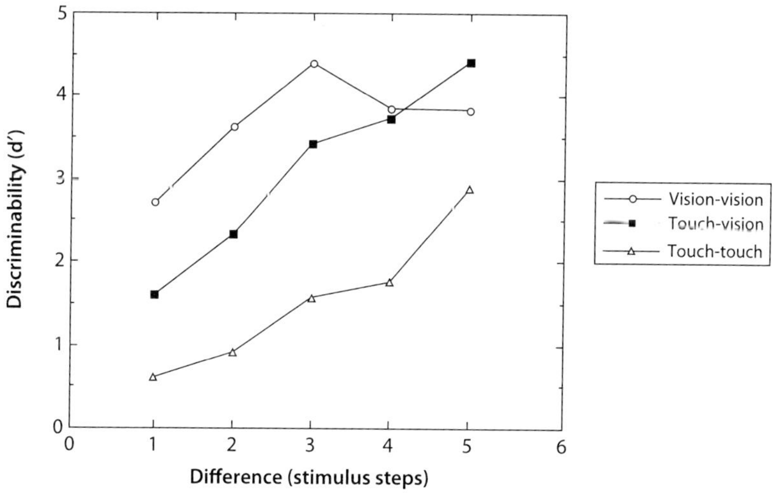

Figure 4.6. Discriminability (d') of curvature derived from same/different judgments, as a function of the stimulus difference (in stimulus steps), for visual-visual, tactile-visual and tactile-tactile comparisons

Reproduced with permission by publisher from Ittyerah, M. & Marks, L.E. (2008) Intra-modal and cross-modal discrimination of curvature: Haptic touch versus vision. Current Letters in Psychology, 24, 1–11. Copyright © Current Psychology Letters, 2008

In general, the modalities order themselves consistently, with visual performance (V–V) again best and haptic performance (T–T) poorest. Over the smallest stimulus differences (steps 1 to 3) where the measures of d' are least susceptible to fluctuations in extreme probabilities and therefore most reliable, the values of d' average 3.5, 2.2, and 1.0 in conditions V–V, T–V, and T–T, respectively. The absolute values of d' reflect, of course, the particular values of curvatures used in the experiment. Nevertheless, the way these values vary across modalities suggest that, for any given pair of stimulus objects, unimodal visual sensitivity to small differences in curvature is superior to cross-modal sensitivity, which in turn is superior to unimodal haptic sensitivity. Perhaps haptic performance is affected by the way the fingers moved over the curved surfaces. It may be relatively easier for the left and right index fingers to move in opposite directions than for both to move in the same direction (Kappers & Koenderink 1996). It is not clear, however, whether a preference for movement direction would affect sensitivity to differences in curvature.

Figure 4.6 shows that the accuracy of responses to same pairs of stimuli (step 0) is much smaller with haptic comparison than with intramodal visual or with cross-modal comparison. And accuracy of responses to different pairs is also smallest, by and large, with intramodal haptic comparison. As Figure 4.6 shows, over the three smallest physical differences, where the measures of d prime are most reliable and least susceptible to variability associated with extreme proportions, unimodal visual performance exceeds cross-modal performance by about one d prime unit, essentially, one standard deviation unit, and cross-modal performance similarly exceeds unimodal haptic performance by about one d prime unit. Thus the modalities differ in their sensitivity to detection and discrimination thresholds. Among sighted subjects vision is at least two d primes (standard deviations) above that of touch for discriminating curvatures of objects. However in blind conditions the modes of perception differ from that of sighted conditions and knowledge of these conditions is valuable for understanding tactile processes.

4.10 Do the hands differ in haptic cognition?

Shapes and sizes of objects are felt and known by the movement of the hand/s over the object. This is more often in blind than in sighted conditions where the accuracy of the features of the perceived object is better known with three dimensional than two dimensional information. The question as to whether there are differences between the hands for haptic spatial coding of nonverbal tasks is of concern because hand actions are predominantly controlled by the contralateral hemisphere (Brinkman & Kuypers 1973). Since evidence indicates that active touch is superior to passive touch (Heller & Myers 1983; Katz 1989), it is of relevance to briefly compare behavioural findings on active touch, which involve finger and hand movements, compared to those on passive touch, which does not. There is evidence that movements are mainly processed in the left hemisphere (De Renzi, Faglioni, Lodesani & Vecchi 1983; Kimura 1993). When a clinical subject was passively guided to explore a stimulus, there was a left hand/right hemisphere advantage for remembering locations. Movements have been found to involve the left hemisphere, though right hemisphere posterior parietal and inferior intra parietal sulcus regions dominate in visuospatial orienting responses (De Renzi 1978, 1982). Studies of brain lesions and functional magnetic resonance imaging (fmri), PET (Positron emission tomography), and TMS (trans cranial magnetic stimulation) have shown that attention to movement involves the posterior parietal cortex more in the left than the right cerebral hemisphere (Rushworth, Nixon, Renowden, Wade & Passingham 1997). A PET study suggested that left hemisphere prefrontal, premotor and parietal regions were dominant for action in general, irrespective of the hand used (Schluter, Krams, Rushworth & Passingham 2001). Such findings might suggest that the right hand performs better for active touch and the left hand performs better for passive touch. There is some evidence to that effect. Lechelt (1982) and Smith, Chu and Edmonton (1977) found that the left hand was better for passive identification of the location of stimulus characteristics, such as dots in a matrix and the right hand was superior for the same spatial task when active scanning was used. Alternately it can be inferred that the left hand is more sensitive and has a lower threshold than that of the right hand (Delantonio & Riggio 1981; Riggio, Dellatonio & Barbato 1980). There is also evidence that spatial tests that test standard spatial ability such as spatial rotation, spatial inference and reasoning tasks also require logical inference and reasoning. Performance on such standard tests of spatial ability was impaired in sighted subjects with lesions in the left and the right cerebral hemispheres (Langdon & Warrington 2000). Millar (2008) argued that behavioural investigations that interpret a left hand advantage as evidence for spatial coding must consider what factors, other than purely tactile inputs may be involved in given spatial tasks. The discrepancies in findings that relate better processing with the left or right hand to spatial coding suggested the need for a test of coding in active touch that would test spatial coding and hand effects independently.

Although brain- behaviour relations are not well understood in infancy and we do not know if the somesthetic pathways have an adult- like crossed pattern, it is quite conceivable that studies of the development of handedness could reveal at least some aspects of these associations (Michel 1988). There is recent evidence

(Roeder et al. 2008) that performance differences between the left and the right hands decrease with age indicating maturity of fibres in the corpus callosum that cross over from one cerebral hemisphere to the contralateral side to control movement on that side. Performance differences between the hands for simple tasks such as tapping were evident from ages 5 to 8 years and gradually declined and became near equal by the ages of 14 years. Therefore the equipotentiality between the hands was evident during early adolescence. This is not ambidexterity when certain actions are performed better by one hand such as writing with the right hand or holding a racquet for hitting a ball with the left hand. Rather, the potential ability of each hand is evident when performance is possible with the preferred and the non preferred hand. This has been observed not only in two handed performance as in juggling objects or piano playing but also in the ability of each hand to subitize items equally well (Plaiser, Bergmann Tiest & Kappers 2010). This indicates that the potential ability of each hand for tactile enumeration is not confined to the preferred hand.

4.11 Millar's reference hypothesis

The tactile perception of an object is more accurate with systematic than unsystematic exploration. Accurate haptic coding of information is dependent upon reference frames. The importance of reference frames for accurate coding of movements was emphasized by Jeannerod (1988), Paillard (1991) and Berthoz (1991). Systematic exploration of stimulus characteristics with the hand or fingers requires an anchor or reference point that can be recognized as the end and starting point of an exploratory movement. To know about spatial processes independent of hand effects, Millar and Al-Attar (2003b) tested two hypotheses. The first hypothesis that the left hand is better for spatial tasks, predicts a left hand advantage for performance in all conditions. The alternate reference hypothesis predicts significantly greater accuracy in haptic recall with additional external reference information than in conditions that do not provide additional reference information. The reference hypothesis assumes that distance and location judgments are spatial tasks. Haptic distance judgments are not solely kinesthetic inputs and movement distances can be coded spatially if they are related to reference information (Millar 2008). Millar and Al-Attar (2003a) found that haptic distance judgments do involve spatial coding. Recall of a repeated small distance was disturbed not only by a movement task, but also by a spatial task that required no movements. In a subsequent study Millar and Al-Attar (2003b) required subjects to recall distance or locations of hapically felt extents. The control condition consisted of scanning the critical distances or locations in presentation and recall without touching any other part of the display or surround. In the experimental or reference conditions, subjects were instructed to use an actual external frame around the stimuli, and also their body midline for reference. The results showed that the added reference information reduced errors very significantly compared to the normal conditions, regardless of whether the left hand scanned the distance in control and frame conditions and right hand was used for the frame, or whether the right hand scanned the distance in control and frame conditions and the left hand was used for the frame. The left and right hands did not differ from each other in accuracy in either control conditions or in reference instruction conditions. The results supported the hypothesis that the use of external frame and body centred reference cues make haptic distance judgments more accurate (Millar & Al-Attar 2004). Accuracy of recall with the left hand did not interact differentially with the increase in accuracy with the instructions to use reference cues. Scanning the distance would involve left hemisphere processing of the movements as well as the spatial aspects of relocating the end position from the new (guided) starting point, and therefore right hemisphere processes also. Cross lateral effects from both right and left hemispheres that inhibit or counterbalance each other would explain why the left hand did not perform better than the right and why it did not relate differentially to the advantage in accuracy from instructions to use spatial reference cues. The important finding was that instructions to use body centred and external frame cues for reference improved recall accuracy for both distance and locations, independently of hand performance, task differences and movement effects. The increase in accuracy of recall for both distance and locations with the use of reference cues demonstrated the validity of operational criteria for spatial coding implied in the reference hypothesis (Millar & Al-Attar 2003a, 2003b). Thus reference information can be used as a reliable test of spatial coding.

Millar and Al-Attar (2004) further tested how egocentric and allocentric coding relate to each other. The hypothesis that haptic targets can only be coded spatially in relation to body centred cues would predict that providing haptic cues explicitly from an external surround would not improve recall accuracy beyond that with only body centred reference cues. Millar and Al-Attar reasoned that if on the other hand the difference in spatial coding was due solely to the lack of external reference information that is normally available in haptic task conditions, providing external haptic cues explicitly for reference in a spatial task should improve recall significantly.

Millar and Al-Attar tested subjects with a spatial task that people might actually encounter in daily living. The task was to remember the precise location of five shape symbols as landmarks that had been positioned randomly as raised symbols along an irregular, but easily felt raised line route. This map like layout had an actual tangible rectangular surrounding frame. Each subject was presented with the map like layout placed on the table and aligned to the subject's body midline. The subjects placed the fingertip of their preferred right hand at the start of the route and scanned the route from left to right in all presentation conditions and briefly stopped on each landmark symbol they encountered on the route, in order to remember the position during recall tests.

There were four reference conditions. In the body aligned control condition, the map was aligned to the participant's body midline in table top space for scanning the landmarks. The test map did not contain the landmark symbols, although it was aligned to subject's body midline. Body centred cues were thus available as reference cues for the locations in presentation and for retrieval in recall. To test body centred coding, the presentation map with landmark symbols was aligned to the body midline of the subject, but the test map without the landmark symbols was rotated clockwise 90 degrees relative to the body midline in the recall test so that body centred coding of the locations was disrupted. The frame surrounding the map was not touched or mentioned at any time, so that there were no external cues in these two conditions. This was compared with accuracy when body centred coding remained intact in recall.

To isolate the use of external reference information, body centred reference was disrupted by map rotation. The subjects were alerted to the external tangible frame and were instructed to feel the frame with the free hand both in presentation and recall and to use information from the frame to specify and remember the location of the felt targets relative to the frame. The fourth reference condition combined intact body centred reference with external reference frame information. The map to body orientation remained the same in recall as in presentation. In addition, the subjects were instructed to use information from the surrounding frame relative to the target locations in presentation and in recall.

The findings showed that disrupting body centred cues by rotation increased errors significantly compared to intact body centred coding in the body aligned condition. The critical findings were a significant decrease in positioning errors with added external reference information when body centred coding was disrupted by rotation, compared to the rotation condition that lacked external reference information. The condition with intact body centred cues and added external reference information was more accurate in comparison to the body aligned condition without external cues, and more accurate also than the condition with added external information, when body centred coding was disturbed by rotation. Accuracy with added external reference information disrupted by body centred coding did not differ from intact body centred coding without external reference information. Separating and combining external and body centred reference showed that external reference cues can also be used with purely haptic information and this seemed to be as equally effective for spatial coding as was body centred reference information.

4.12 Is vision necessary for haptic perception?

The notion that vision has a pervasive effect implies that vision will produce greater accuracy in spatial tasks such as remembering positions or locations even when th,e visual condition excludes visuo-spatial information. By contrast the reference hypothesis (Millar 2008) implies that visual information benefits performance only to the extent that it provides cues that are relevant to a task. Millar and Al-Attar (2005) tested whether the mere presence of vision can enhance haptic performance in a spatial task. They combined touch with three forms of reduced vision such as diffuse light perception, peripheral vision and tunnel vision. In diffuse light perception vision provided light but no shape or shadow vision or information about finger movements or visuo spatial cues of any kind. In peripheral vision there was blurred vision of a relatively large space outside the central area. Tunnel vision excluded peripheral vision. Looking directly at a shape provided clear focal vision in a small area of about 3.7cm diameter.

The test required the subject to scan a tactile map with location along a route that could be felt by the scanning finger and recall the felt locations on a test map that was identical to the tactile map, but without the landmark symbols for location. The subject was required to identify the felt locations on the test map by stopping at the appropriate location. The experimental condition included full vision, peripheral vision, tunnel vision, diffuse light perception and total blindness. The conditions of residual vision and total blindness were simulated with designed spectacles. The results showed that full vision with touch produced the largest improvement compared to touch alone than any other condition. Peripheral and tunnel vision added to touch were less accurate than full vision, but they were more accurate than either diffuse light perception or touch alone. The important finding that supported the reference hypothesis as against the visual information hypothesis was the effect on touch of diffuse light perception, which provided vision but excluded spatial cues. Diffuse light perception was the only form of reduced vision that did not improve haptic perception. The finding supported Millar and Al-Attar's hypothesis that vision improves haptic performance only in so far as it provides relevant cues.

Haptic perception of form takes place largely by inspection of shapes by the palm and fingers of the hand as they move over the surface of an object. The property of shape or form refers to the spatial layout of the object, which may be specified in terms of contours. An isolated contour may itself be described with respect to its extent and orientation, and interactions among contours, or patterns, may in turn be described in terms of component contours or as combinations of angles, straight or curved surfaces, and other distinctive features. The shapes of many objects may be adequately described in terms of angles and curvatures of various degrees and proportion. In haptic exploration of shapes, the fingers move over the various angles or bend over the curves.

Studies with infants have revealed that recognition of objects either seen or held in the hand is possible during early periods of infancy (Hatwell 2003). Streri (1987) and Streri and Pecheux (1986) demonstrated that infants can habituate to the shape and size of objects held in their hands, showing evidence of perceptual ability at an early age. In the habituation procedure, objects were presented in the infant's hand on several consecutive trials and a decreasing duration of grasp was observed over the trials. Habituation in infancy involves complex processes such as recognition memory and information processing in vision (Cohen 1998, Olson & Sherman 1983) and haptics (Streri 1993). Streri and Gouarir (1996) found that six months old infants hold an object longer with the right than the left hand. If habituation in infancy is an index of cognitive processing (Cohen 1998), then this finding indicates that the right hand (left brain) takes in information about object shape and stores and remembers the information even if there is a left hand distraction. Since the holding times were longer for the right than the left hand, Streri and Gouarir (1996) suggested that the holding times for the right hand revealed a motor ability, controlled by the left hemisphere and the holding times of the left hand revealed a perceptual ability, controlled by the right hemisphere. The left hand seemed to be more sensitive to a change of shape during the test trials. After familiarization to an object under visual control, the infants' left hands discriminated between different object shapes whereas their right hands did not. A left sided sensory bias seemed to exist.

The haptic system involves two functional characteristics simultaneously: it processes the spatial properties and carries out this task in a sequential manner (Hatwell 1994, Hatwell, Steri & Gentaz 2003). These functional properties make its study complex and few studies have investigated the information processing of object shape to shed light on the relationship between haptic perception, handedness and hemispheric differences in infancy. Streri (2002) studied differences between the global and analytic aspects of haptic form perception in 4 month old infants. Habituation and discrimination of the shapes were conducted with the contour of objects, the features remaining identical (faces cut out or embossed on a card board shape, either curvilinear or rectilinear in shape). The analysis revealed that in the global condition infants discriminated the contours with their left hand, but not with their right hand. In the analytic condition infants discriminated the details with their right hand, but not with their left hand. Thus the hands seem to specialize early in infancy in information processing according to the analytic/global mode.

The ability to transfer shape information from one hand to the eyes has been demonstrated in human newborns (Streri & Gentaz 2003). It is important to know the limits of this ability because it does not support neither the traditional (Molyneux question; Berkley 1996; Hume 1960; Locke 1975) nor modern connectionist models (Elman et al. 1996; McClelland, Rumelhart and Group 1986) notions, that sensory modalities cannot communicate in the newborn, and need to be associated by experience and learning in order to be coordinated. Handedness in cross-modal transfer from touch to vision investigated in newborns revealed intersensory integration soon after birth. Streri, Lhote and Dutilleul (2000) found that newborns can discriminate intra-modally with each hand between two object shapes (prism or cylinder). Using an intersensory paired preference procedure, Streri and Gentaz (2003) observed that newborns can visually recognize the shape of an object that they have previously manipulated with their right hand, out of sight. However, this ability was not present for the left hand. Streri and Gentaz's (2003) study revealed that cross-modal recognition of shape from the hand to eyes in newborns depends on the hand stimulated during the tactual habituation phase. Asymmetries favouring the right side (Streri 1993) have also been observed in newborns. In any case the study confirmed an early ability to extract shape information in a tactual format and transfer it to a visual format, independent of common experience.

The tactual perception of objects is mostly sequential. In order to apprehend the whole object, voluntary movements need be made to extract information. The kinesthetic perception resulting from these movements is closely linked to cutaneous perception and together they form active touch or haptic perception. The sequential property of haptic perception increases the demands on working memory to obtain a unified representation of the object (Gibson 1966). The hands are also motor organs used to reach, hold transform or transport objects. Therefore the hands are a complex system that involves two functions: a perceptual function (knowing) and an instrumental function (doing) (Streri, 2005). Before the age of five months, manual exploration though limited, allows the properties of objects to be apprehended. After the age of five months, the emergence of instrumental function in the visual/ prehensile coordination replaces the first forms of tactile interaction with objects. The manual system now performs its transporting and exploration functions under visual control. Thus the status of the hands is modified (Streri 2005).

The haptic modality emerges early in ontogeny and is the first modality to be expressed in the fetus. In newborns, grasping and avoiding reflexes were regarded as the dominant behaviours. During the first five months, these reflexes were considered as a prelude to more voluntary actions, such as holding and releasing objects. Besides simple hand reflexes, infants at birth were able to handle small objects. Streri, Lhote and Dutilleul (2000) used the habituation/dishabituation procedure (Streri & Pecheux 1986) and found that neonates (the youngest being

16 hours old) were able to detect differences in the contours of two small objects (a prism and a cylinder) with both the right and left hands. The habituation/dishabituation procedure involved two phases. During the habituation phase, the infants received a small object in their hand. A trial began when the infant held the object and ended when the infant dropped the object or at a later duration defined by the experimenter. This was repeated several times. Over the course of these trials, a decrease in holding times was observed. In the dishabituation phase a novel object was put in the infant's hand. If an increase in the holding time of the novel object was observed, it was inferred that a reaction to novelty was revealed and that the infant was able to discriminate between the two objects. Moreover it is now well established that process of habituation and memory are connected to reveal instances of representation of the stimulus. Streri et al. (2000) report a first evidence of habituation and reaction to novelty observed for the left and right hands in neonates. From this it can be inferred that haptic memory in its manual mode is present at birth. Geometric shapes, curvilinear or rectilinear, full shapes of objects or objects with holes are well differentiated from the age of two months (Streri 1987; Streri & Milhet 1988). At this age the infant is also able to discriminate a volumetric object (a cotton reel) from a flat object (a cross) (Streri & Molina 1993). However, these responses do not necessarily indicate that the baby has a clear representation of what is being held in the hand. Shape perception may be confined to the detection of features such as points on the object, curves, the presence or absence of a hole, etc. The information is gathered from the exploration of the object enclosed in the hand.

Nevertheless, both hands do not have the same competencies to detect features in complex objects. For example, Pineau and Streri (1990) found that five months old infants were able to detect an unlikely element in a complex object with their right hand. This showed that infants were able to process the contour of an object as well as the details in the object. Streri (2002) demonstrated that in global and analytical tests of objects held in the hands of infants, the infants seemed to prefer the right hand for discriminating features and the left hand for contours. Thus, when an object was held by the infant, a manual specialization was observed indicating hemispheric specialization very early in infancy. When two handed exploration was initiated from the age of four months, the infant could hold a voluminous or large sized object. Infants have also been found to be capable of detecting differences in the weights of two objects (13g or 47g) held in each hand (Streri 1993). In summary, when the haptic system is tested, its information gathering and processing capacities about object properties are relatively good. They do not depend on the specific exploratory procedures identified in adults (Lederman & Klatzky 1987) as suggested by Bushnell and Bordreau's (1991) analysis and indicated by Millar (1994).

Haptic touch is a complex perceptual system, relying not only on several low-level sensory mechanisms (inputs from mechanoreceptors, kinesthesis, and proprioception), but also on higher-level inputs that depend on the type, size, meaning, and familiarity of the stimulus objects, as well as the demands of the task (Millar 1999). Although vision may be specialized to process information about spatial layouts, and hearing specialized to process information about temporal patterns as in speech, touch can convey both types of information. Further, haptic perception depends on the interaction between sensory input and motor behavior. The interaction between input and output systems, linking, for example, the size and shape of an object perceived haptically to the direction and extent of the hand's movement over the object is important in determining the type of information available and how the information is processed.

The senses convey information from more than one stimulus or object. Just as vision provides information from multiple stimuli such as items laid on a table, the haptic system as well provides information from more than one stimulus or object. During search periods, salient features have been found to 'pop out' from a background of other stimuli. Treisman (1985) observed that the pop out of salient stimulus characteristics such as a red X in the midst of green Xs involves a preattentive state of attention and does not require any effort by the sensory system. Similarly stimulus characteristic have been reported to pop out during tactile searches, indicating a preattentive sensory process in the tactile sensory system. Plaisier, Bergmann and Kappers (2008) found that items pop out during touch with a single hand sweep. When the target item was a rough sand paper shape in the midst of fine sand paper stimuli that served as distracter items, the target popped out with a single hand sweep indicating that the distracter items have little influence during the search. However searching for a medium rough target among fine distracters was slower and required more attention.

Recognition of tactile characteristics is not only confined to surface properties such as texture but also involves contour information such as sharp edges or angles about the stimulus. Plaisier, Bergman and Kappers (2009a) investigated the saliency of three dimensional shapes in a haptic search task. When the target was a cube amongst distracters that were spheres, the cube was detected as fast as 25 milliseconds in the midst of spheres because edges and vertices were the salient features. However it took longer to detect a sphere among cubes. The reason for this could be because cubes have a larger contact area than spheres and give rise to larger frictional forces as compared to spheres that may slide away easily.

Enumeration has been considered to be a visual process even if the items are held in the hand during simultaneous visual inspection. It is known from visual studies that people judge the number of items in a display rapidly and errorlessly up to a certain number usually three. This phenomenon is known as subitizing

(Kaufman et al. 1949). In subsequent trials with increasing number of items, reaction times and error rates increase as a function of set size. Subitizing is characterized by a sharp upward bend in the slope of the reaction times and error rates with increasing set size. In vision the location of the bend is at about four items (Mandler & Shebo 1982) and indicates the item content when counting begins. Therefore enumeration involves a fast subitizing process and a slower counting process when items are more than three in a set size. Evidence indicates the involvement of separate neural mechanisms for the two processes. Findings showed a left hemispheric specialization for subitizing and a right hemispheric mechanism for counting (Pasini & Tessari 2001; Piazza. Mechelli, Price & Butterworth 2006). Plasier, Bergmann Tiest and Kappers (2009b) found subitizing in touch to be not different from that of previous studies in vision. Subjects grasped one or more spheres with their preferred hand to make numerosity estimations of the spheres in their hand. The typical upward bend in the slopes of the reaction times was obtained when subjects grasped more than three items indicating the bifurcation of the slope where subitizing terminates and the onset of counting begins.

Plaisier et al. (2009b) also observed that numerosity estimations as subitizing does not vary with constant relative spacing of stimuli. Subtitizing was obtained with small number of spheres with constant relative spacing as 1/2/4/8 and 16 items as it was obtained with constant absolute spacing of items such as 1, 2, 3, 4, 5, 6 and 7 spheres. Therefore the differential spacing or grouping of items does not affect subitizing. Furthermore an absence of numerosity information such as varying sizes of spheres with constant relative spacing did not evoke subitizing. The error rates were higher in distinguishing the first from the second sphere in contrast to the fast enumeration of small set sizes during subitizing. Subitizing in touch is a fast process of errorless enumeration as in vision.

Plaisier, Bergmann Tiest and Kappers (2010a) continued to demonstrate that subitizing items can be performed equally well with either hand. They showed that the left and the right hands were able to subitize three items. Furthermore during bimanual trials when items were presented in clusters to each hand, clustering enabled subitizing up to six items and reduced response times considerably for larger numerosities as compared to unimanual presentations. This indicates that enumeration of numerosities from the counting range is performed by a combination of subitizing and counting showing that subitizing is a subprocess of counting and is independent of the sensory modality.

When objects are laid out on a surface like a table or if they are kept together in a basket, we need to haptically identify and distinguish one object from another. To be able to know which part belongs to one object or to another is known as item individuation and is an important aspect of haptic perception. There may be instances when we grasp several objects together such as a bunch of keys or varying sizes of beads and try to select a particular key or bead with our fingers. The question of interest was how items that are grasped together are individuated. Plaisier, Bergmann Tiest and Kappers (2010b) demonstrated that the quantity of items that were haptically subitized did not differ with size or shape. Even when items were reduced in size, the subitizing quantum did not go beyond three items though the number of items that could be grasped may have increased. Similarly increasing item size by varying the mass and volume of the items did not affect the subitizing quantum of three items. The subitizing range was not affected when heterogenous shapes were presented such as cubes and spheres. Therefore the individuation mechanism that is available for subitizing to occur does not rely on size or shape features of the object. Though the subitized content of three is obtained haptically, the content of the items as to whether there are two spheres and a cube may not be known. However Plaisier et al. only tested for enumeration and did not investigate categorization that may have explained how many cubes or spheres were available for inspection in the hand. It is evident that distinctive features of the object need to be isolated to tell a cube from a triangle or a circle from a sphere. The finding that the range of subitizing items does not differ between the hands indicates that haptic features of object recognition such as shape, material and size (Plaisier et al. 2009a) can be processed equally well with the left or the right hand in the absence of vision.

Studies have consistently revealed that the processing of touch and vision differ in that touch is processed serially while attending to information on all sides of the object as the hand and fingers feel the contour and its features, whereas vision is typically processed globally and often the back portion or the third dimension of the object is not visible. Theurel, Frileux, Hatwell, and Gentaz (2012) indicated that processes of categorization of objects such as the formation of prototypes are characteristics of the visual modality. Vision provides a wide range of information simultaneously from near and far space. This enables the learning of prototypes of objects as is typically characteristic of vision. However in touch such prototypical information is not learnt and subjects respond to stimulus characteristics that differentiate one stimulus from another as in item individuation (Plaisier et al. 2010b). Stimulus characteristics such as corners of shapes or spherical extents are detected by the fingers or the palms of the hand as it touches and moves over the entire object without generating the circumstances for prototype learning unlike vision that provides frontal information of the whole object to generate typical features of prototypes. Thus haptic cognition is distinct from vision and other sensory modalities.

It is now reasonable to propose a theory of equal hand ability in the chapter to follow given the substantial evidence for recognition of objects with each hand early in infancy, as well as the equal enumeration ability for each hand to subitize information. Furthermore each hand controlled by the contralateral hemisphere has shown specific processing ability for global or local characteristics of the stimulus or for movement or location indicating the complementary functions of each hand in stimulus perception. The evidence of equal hand ability in blind conditions enabled by body centred and external frames of reference during spatial coding of map like information is also good evidence that the hands do not differ in ability. In the forthcoming chapter an attempt is made to reveal the potential ability of each hand from studies in blind and sighted conditions to clarify that even if one has a hand preference, each hand may be as able as the other in task performance.

CHAPTER 5

Hand and skill

Toward a theory of equal hand ability

Hand function and manual skills have been important in primate evolution and have a central role in human ontogeny. Common activities involve the hand for holding and manipulating objects. Though most actions have a complex pattern of activity that involves the whole of the upper limb and trunk, manipulation depends upon the organization and coordination of movements of the digits. Manipulation is concerned with the handling of objects to achieve particular goals and dexterous manipulation refers to being adroit or nimble to perform skillfully. Typically performance has been assessed in terms of speed and accuracy, though little is known of how tasks are performed or of patterns of digit coordination during skilled actions, despite their relevance to understanding early motor development (Manoel & Connolly 1997).

In population studies on lateral asymmetries, stable predictions about the lateral dispositions of populations have been made and genetic models constructed (Annett 1985; McManus 1985). Handedness is assessed as either preference or skill (McManus 1991). Measures of handedness/preference usually show a bimodal distribution, with few subjects appearing ambidextrous (McManus et al. 1988). Handedness skill measures are more inconsistent: some tasks such as the Annett pegboard (Annett 1970b) give a unimodal distribution, suggesting a continuum of handedness; whereas other tasks such as putting dots in a circle (Tapley & Bryden 1985) or squares (McManus 1985), give a bimodal distribution, implying two discrete categories of individuals, analogous to the two modes found in handedness/preference measures.

However when laterality in the motor control of individuals was examined, there was little consensus. There was also a gradual change in the relation between asymmetry in underlying structure and behaviour (Kertesz & Geshwind 1971) that is observed as organisms become increasingly complex. According to Peters (1991b) if behavioural asymmetries (motor asymmetries) have an adaptive function, then in order to be enabling in certain circumstances, such motor asymmetries must not be entirely controlled by the underlying structural factors. If it were so, an organism would be limited in responding to the external world. For example, if an animal would prefer to reach for insects with its left paw, it would be quite maladaptive if it could reach only with its left paw and not use the right paw. Galaburda (1991) indicated that a larger lateral corticospinal tract on the right side supports a right side preference, but does not preclude a high degree of motor control on the left side. Peters (1991b) points out that the relation between hand preference and skill and the larger pyramidal tract size on the right is not understood. Furthermore speculation had been rife since Kertesz and Geshwind (1971) observed the pattern of cross over of the pyramidal tracts that was characteristic of right handers (left sided tract crossing over before the right sided tract) in all their left handers as well. The minority of cases with an opposite cross over pattern were right handers.

Whatever the relation between handedness and anatomy of the motor tracts, there is evidence that the skeletal structures involved in actions are not as limiting in terms of behaviour. For example high levels of motor skills can be attained with systems that are not expected to control fine motor activities. Many a handicapped person was known to substitute the hands with the feet or the mouth in skilled performance. These individuals show that movement intent can be expressed flexibly, at an appreciable level of accomplishment, through different executing outlets. Madame Hanakawa from Japan, born without arms was able to produce origami paper folding creations with her feet. The mandolin was skillfully played with the feet by a Frenchman Jean de Hanau and a Miss Sarah Biffin in 19th century Britain, lacking arms and feet produced fine lithographs by holding the implements in her mouth (Jay 1986). If the feet can perform delicate manipulative acts, the hands with their many degrees of freedom are considered to be more skilled than the feet. Therefore differences in performance between the hands are not expected to be greater than differences in performance between the feet (Peters 1991).

Quite recently Funk (2006) demonstrated the discovery of changes in body representation due to a congenital absence of a body part. There is convincing evidence for plastic changes in the motor cortex from brain imaging and behavioural explorations showing the ability of the brain to reorganize neural pathways based on new experiences. The plastic organization of the neural system not only facilitates the migration of neural networks but also serves for functional compensation. According to Funk, in cases of congenital absence of one hand, the deafferented cortical space dedicated for normal limbed subjects to the motor control of hands was not inactive, since the adjacent orofacial motor area had invaded the deaffrenated hand area. Therefore the motor cortex responded to the congenital absence of limbs by reorganizing neural pathways. When the motor cortex of amelics (people with congenital absence of limbs) responds appropriately to actions in the absence of hands or arms, it indicates that they have a normal body schema representing both the missing and the existing hand (Funk & Brugger, 2002).

Phantom limb sensations of people with missing limbs can affect the observers own body representation or schema. The visual analyses of other's actions have been found to influence the body schema of amelics that experience phantom limb sensations. Funk, Shiffar and Brugger (2005) reported that one of the two aplasic persons (without arms or legs) who experienced phantom sensations perceived apparent motions of hand movements, whereas the other who did not experience phantom limb sensations perceived no apparent hand motions. Therefore phantom experiences can affect the visual analysis of the human body as do years of sensory motor experience in normal limbed people. There is also evidence of motor imagery aiding the rotation of the missing hand among people with hand amelia. Funk and Brugger (2008) compared motor imagery performance between normally limbed subjects and subjects with one or both hands missing in a hand rotation task. The subjects were required to rotate pictures of hands depicted in different degrees of rotation as either a left or a right hand. Subjects with either a left or right missing hand had slower response times to hands corresponding to their missing hand as compared to times for their existing hand. Amelics with both hands missing showed prolonged reaction time to any hand depicted in an unnatural orientation such as a medial display as against a lateral display, particularly if they experienced phantom sensations of the absent hand. The visual recognition of a hand that has never been physically present is prolonged. Nevertheless these examples of cerebral plasticity show that movement intent can be expressed flexibly even in the absence of limbs.

The other line of evidence concerns the prevalence of left handedness. Handedness is sensitive to cultural pressures. Prevalence of left handed writing has been reported from less than one percent (Komai & Fukuoka 1934) to more than 13% (Spiegler & Yeni-Komshian 1983). Gender differences in prevalence range from samples where no female left handed writers were observed (grade 8 sample of Komai & Fukuoka 1934) to those where females outnumber male left handed writers (Levander & Schalling 1988). Caliskan and Dane (2009) reported more left handers in blind than sighted children suggesting the importance of vision in the development of normal typical cerebral lateralization or hand preferences. Their survey showed that the incidence of left handers were more among males than females both for the blind (18.23% in boys; 17.02% in girls) and the sighted children (11, 02% in boys; 7.52% in girls). However in India with large numbers of blind children, the incidence of left handers has been quite similar in normal blind and sighted children. Left hand preferences for a series of actions including writing have shown a range from 4 to 7 percent in congenitally blind and sighted children (Ittyerah 1993, 2000, 2009). According to Peters (1991) evidence is likely to indicate that studies of asymmetries in motor performance will show an effect of habit on the executing motor systems. Developmental studies on human infants (e.g. Bresson, Maury, Pieraut- Le Bonniec & de Schonen 1977; Lynch, Lee, Bhat & Gallaway 2008) show that asymmetries in movement occur before the motor systems for which an anatomical asymmetry has been documented are fully functional, such as asymmetries in simple orienting or reaching responses. Peters (1991b, 1987) suggested that an early lateral bias in orientation is transformed into an attention bias that becomes important in skilled movement. The crucial point is that the attentional process, although normally concordant with structural asymmetries, remains superordinate in the level of execution, and is not therefore wholly constrained by demonstrated structural motor asymmetries (Nathan et al. 1990; Tan 1985). Thus while one is inclined to focus attention on the preferred hand, one can readily switch attention to the other hand when changing hands in unimanual activities.

Peters (1990a) argued that in bimanual activities, the hands assume different roles and the allocation of attention reflects it. There are two important aspects: first when hands become specialized in unimanual activities the switching of hand roles requires merely that the other hand also masters the skill. In bimanual tasks both hands have to relearn. Therefore there is greater incentive for the consistency of hand roles in bimanual than in unimanual tasks. Second, the attentional disposition across the two hands harmonizes with the division of motor roles. While some switching of attention is possible, fast concurrent movement of the two hands will often demand a split into primary focused attention toward the preferred hand and subsidiary attention to ward the non-preferred hand (Peters 1990a). Although the preferred hand often takes the more demanding role, this is not always so, and the assignment of roles follow attentional preferences more than motor demands. For example while playing the piano the motor demands of the right hand that plays the treble clef are not more than that of the left hand that plays the base clef. When the motor demands of the two hands are kept identical and the allocation of attention is manipulated, it becomes clear that performance is much better when attention is focused on the preferred hand (Peters 1985). Therefore according to Peters (1991b), lateral biases in attention take precedence over motor biases and are operative before structural biases become functional in the executing motor system.

Peters (1990b) clarified this notion in different groups of left handers. Subjects were given a short handedness questionnaire. If they wrote with the right hand and considered themselves right handed, they were called right handed. If they wrote with the left hand and preferred the right hand for no more than one other activity, they were called consistently left handed. If they wrote with the left hand and preferred the right hand for two or more of the activities, they were called inconsistently left handed. The inconsistency referred to the choice of different hands for different activities, rather than the flexible use of either hand for a given activity. When hand strength was measured, it was found that right handers had a stronger right hand. Left handers showed no significant differences between the two hands. However, when the left handers were subdivided into consistent and inconsistent subgroups, they showed strength asymmetries in opposite directions. In a throwing task, the left handers as a group showed no significant hand asymmetries, but when subclassification was used the two groups of left handers showed asymmetries in opposing directions. The left handed superiority for the consistent left handers was observed, but the inconsistent left handers as a group, were stronger in the right hand and threw more accurately with the right hand. When inconsistent left handers were tested on finger tapping speed they were faster with the left hand. Thus in this group, the direction of lateral specialization was a function of the particular motor task. Inconsistent left handers also performed better with the left hand on another test of fine motor skill, the Perdue Pegboard task. Thus, inconsistent left handers showed a right side specialization for strength and ballistic activities of the arm, and a left side specialization for fine manual activities.

According to the attentional hypothesis, in right handers and consistent left handers, lateral specializations for strength and skill are congruent and attentional specialization need not be considered separately. Inconsistent left handers focus attention to the left hand for writing and fine manual skills but to the right side for throwing and kicking. In bimanual tasks as in finger tapping, no significant asymmetry differences were observed between the consistent and inconsistent left handers. Peters hypothesized that the differences between the consistent and inconsistent left handers may be in the development of motor systems that control the axial and distal musculatures. Steinhuis and Bryden (1989, 1990) have for example shown that hand actions may be controlled by the musculatures that are involved in performance. Actions such as picking up objects may be controlled by the distal musculatures, whereas actions such as batting may be controlled by the axial musculatures of the hand and back. While lateral specializations in these two systems develop to be laterally congruent in most individuals, it is possible that such specializations are not laterally congruent in a small proportion of inconsistent left handers. Performance in this group of left handers indicates that hand skills are not confined to the preferred hand. Recently Buckingham Main and Careey (2011) compared left and right handers in a cued bimanual reaching task and found asymmetries in motor attention. The left and right handers received a tactile cue prior to bimanual reach to target pairs. Right handers took longer to inhibit their right hand responses and committed more right hand errors, suggesting that their dominant hand was primed to move than their nondominant hand due to attentional biases to the right hand (Peters 1991). The left handers showed neither of these asymmetries, suggesting they lack an equivalent dominant hand attentional bias.

5.1 Developing motor skills

Studies of neurogenesis in birds initiated interest in the study of neurogenesis in animals and provided the impetus to investigate neurogenesis in adult mammals (Johnston & Rose 2002). The left and right eye systems of chicks have been found to respond to different characteristics of stimuli. Using different experimental training techniques such as imprinting or passive avoidance learning, the left eye system was able to differentiate between items like a familiar and an unfamiliar chick, whereas chicks using the right eye system approached both stimuli equally as often (Vallortigara & Andrew 1991). Imprinting is learning to associate a person or object with a particular action and occurs soon after birth as among chicks following the mother hen. Imprinting serves as a social reference for an action or behaviour. Passive avoidance is learning to avoid a noxious stimulus usually with training so that for example, birds are reinforced not to peck at a stimulus. Horn, Bradley and Bateson (1979) localized imprinting related changes in RNA synthesis to a specific brain area in the chick known as the intermediate medial hyperstraitum ventral (IMHV). The IMHV is not a primary sensory area but functions as an association cortex in the avian brain receiving secondary inputs from other sensory regions. The differential functions of cerebral hemispheres in chicks seem to depend in part on the integrity of both hemispheres and on some form of interhemispheric communication between the left and right IMHV (Horn 1985; McCabe 1991). The integrity of the left and right hemispheres in chicks during imprinting or other learning is an analogy of the integrity and interdependence of the cerebral hemispheres that control hand actions in humans.

The evidence on the neurological basis of developing motor skills in humans has indicated a relationship between neuromotor maturity and handedness (Bishop 1990a). Consistent hand preferences were related to an advanced developmental status. In six monthly assessments of handedness, Gottfried and Bathhurst (1983) observed that girls with consistent handedness were more advanced in drawing, indicating a positive relationship between handedness and callosal maturity. This raises the question as to whether children with unestablished handedness are less consistent in crossing the body midline or commit more mirror movements and are less able to perform bimanual asymmetric actions.

Palmer (1964) regarded stable hand preference as the outcome of a process of neurological differentiation. The normal infant progresses gradually from a state of undifferentiated bilaterality, with innervation of the trunk developing first, followed by the arms and finally the hands. If this developmental process is slowed or arrested, there would be lack of differentiation and poor motor skill. On a handedness inventory, a completely undifferentiated child would respond at random. In this situation, on a binary handedness classification, left and right handers would be equally common, not because of any factor biasing toward the left side, but simply because the normal bias favouring the right side is absent, showing an increased randomness of hand preference (Palmer 1964).

Soper et al. (1986) hypothesized that ambiguous handedness represents a pathological subtype found in clinical subpopulations with severe bilateral brain damage which prevents the development of manual dominance. This contrasts with Bishop's (1990a) suggestion that ambiguous handedness is an indicator of motor immaturity. If ambiguous handedness corresponds to developmental delay rather than deviance, then the extent to which ambiguous handedness is shown, should be predictable from the level of motor skill. Lonton (1976) tested the relation between ambiguous handedness and brain pathology from drawings and writing in children with spina bifida and hydrocephalus. A substantial minority of the spina bifida children used both hands, a behaviour that was rarely observed in the youngest normal control children. Anderson and Spain (1977) observed stretching and thinning of the corpus callosum in children with spina bifidia. These children showed mirror movements beyond the usual age. Both the increased randomness hypothesis (Bishop 1990a) and the pathological hypothesis (Soper et al. 1986) assume that the excess of non right handedness is caused by non genetic influences. For example in many mentally handicapped children, lack of right handedness does not indicate left handedness or ambidexterity, so much as randomness on hand preference inventories (Bishop 1990a).

Is randomness of hand actions a function of the level of motor skill or pathological phenomena arising as a consequence of callosal dysfunction? Soper et al. found randomness of hand preferences than pathologically enforced left handedness in autistic children. They observed two types of mixed handedness: (1) inconsistent hand actions, and (2) reliable preference for a given activity, though the direction of preference differed from one activity to another. Soper et al. explained this outcome as ambiguous hand actions particularly frequent in autism and those with low intelligence. Autistic children have an absence of hand preference than stable left handedness. Therefore random handedness rather than left handedness is a clinical condition.

Bishop (1990b) found little association between developmental language disorders and non right handedness or severe phonological problems and left handedness in children between the ages of 4 to 8 years. Bishop and Edmundson (1987) observed that slow performance on a peg moving task was correlated with severity of language impairment. As language disorder improved, motor performance improved. Poor peg moving is an indication of neuromotor immaturity rather than brain injury. Motor function appears to be immature in language impaired children, but there is little evidence of consistent lateralization among these children. For Bishop, it is motor performance rather than language status that is linked to handedness. For both Bishop (1990a, 1990b) and Crow (1998, 2000), an absence of hand preference is an indication of low performance on intelligence or ability tests, though Crow like Annett emphasized a sex linked gene to be correlated to lateralization. Other recent examples of random handedness have been observed in clinical groups of children. Domello, Ro¨nnqvist, Titran, Esseily and Fagard (2009) explored effects of prenatal alcohol exposure on functional lateralization in children with fetal alcohol syndrome. Item tasks measuring preferences of hand, foot, eye, and ear were administered to a sample of 23 children diagnosed with fetal alcohol syndrome (FAS) compared with typically developing (TD) children. In addition, a dichotic listening task was administered to a subsample of 11 children with FAS and a TD group of comparable age, sex and handedness. The children with FAS were characterized by increased nonright handedness compared with TD children. No differences were evident for preferential use of foot, eye, or ear. Moreover, children with FAS displayed more right ear extinctions during dichotic listening relative to TD children, indicating a lack of right ear advantage. The results add to findings of decreased manual asymmetry and less left-lateralized speech perception in children with developmental disorders, and indicate high incidence of callosal abnormalities in alcohol-exposed children. In a subsequent study Domellof, Fagard and Ronnnqvist (2010) used 3D kinematic analysis to understand the motor abnormalities in children with fetal alcohol syndrome. An optoelectronic tracking system registered left and right arm and head movements during a sequential unimanual goal-directed precision task in a sample of FAS and TD children. Children with FAS demonstrated poorer task performance compared with TD children. Additionally, analyses of arm movement kinematics revealed atypical spatio-temporal organization in the children with FAS. In general, they exhibited longer arm movement trajectories at both the proximal and distal level, faster velocities at the proximal level but slower at the distal level, and more segmented distal movements. Children with FAS also showed atypically augmented and fast head movements during task performance. Findings indicate neuromotor deficits and developmental delay in goal directed arm movements because of prenatal alcohol exposure. Random handedness appears to be a clinical condition.

5.2 Hand preferences differ from hand ability

The importance of the study of the human hand was emphasized as early as 1840 in Sir Charles Bell's treatise with its singular message that no serious account of human life can ignore the central importance of the hand. This message is as relevant in the present as it was then. The hand is involved from the beginning in an infant's construction of visuomotor, kinesthetic and haptic representations of the world and objects. This is a profoundly important role for the hand, whose importance in ontogeny cannot be overstated. Greenfield (1991) for example in her review observed children as they worked on simple hand oriented problems such as eating, stacking cubes, playing with spoons or copying geometric patterns of sticks. Greenfield observed that children at different ages differ in their attempts to reproduce the geometric stick designs presented by the experimenter. Though the six year old was not completely successful, the seven year old developed the ability to represent the tester's pattern as her brain generated the movement of her arms and hands that consequently reflected her cognitive architecture. The eleven year old displayed mastery of the task and was able to improvise on the tester's pattern. This outcome signalled a critical advance in cognitive function. Thus hand actions reflect cognitive capacity as is evident in different skills such as piano playing, carpentry, pottery or juggling.

Theories of eye hand coordination have been increasingly supported by evidence that the upright position of the head in young infants between five and eight weeks of age increases their reaching accuracy to match that of infants that are twenty weeks of age (Jeannerod 1997a). Although head position serves as an efficient reference system for directing movement toward extrapersonal space, Fraiberg as early as 1968 was able to demonstrate that the body mid-line was the first place to have subjective reality for blind infants, and subsequent evidence shows that self referent cues are reliable (Stelmach & Larish 1980; Millar 1985b). Such fundamental reaching behaviours were evident in sighted infants as well (Wishart 1978), since early in infancy the hand teaches the eye (Hatwell 1987). Even though vision dominates later in infancy (Hatwell 1987) instances of skilled performance in blind conditions (e.g. blind pianists) is an indication of tactile kinesthetic and auditory feedback in the absence of vision. Furthermore, tasks that involve precision grips with the thumb and fingers may be equally well performed with either hand even in blind conditions. In precision grips, the frictional adaptation of the fingers to objects begins to develop by the age of two years so that the object does not slip from the hand (Forssberg, Eliasson, Kinoshita, Westling & Johansson 1995). Subsequently the ability to lift the object with an appreciable anticipatory control indicates a development in the ability to represent the frictional properties of the object in either hand. Performance after practice of manual skills in sorting, stacking, finger dexterity and peg moving (Minnesota Rate of Manipulation) with each hand in blind conditions (Ittyerah 2009) has demonstrated that this representational ability does not rely on vision and can develop in the total absence of vision. Undoubtedly, visual information increases precision in performance, though the criticality of vision as an omnipotent pervasive factor in hand preference or skill is not evident. Efforts should be directed toward studying the role of the hands in task performance rather than the role of vision in task performance.

However there is a need to clarify that lateralization as indicated by hand preference may be different from hand ability. Annett (2002) demonstrated that left and right hands may not differ in performance in peg moving or similar tasks and the only hand performance differences are between that of men and women, with women being more lateralized than men. Therefore the general laterality does not affect ability.

The literature on hand skills fall into three broad domains of explanation. One view is that the preferred hand is the skilled hand and there is a correlation between hand preference and skill. Another is that handedness is not a unidimensional variable and hand actions may be grouped according to the muscle groups involved in performing the tasks. A third opinion holds that the hands may not differ in skill and each hand has the ability to perform an action. The performance of both hands may be as good as each other and may depend upon conditions of task demands. It is in the interest of this manuscript to examine these differing explanations of hand skills. I conclude that there is equal potential for skill between the hands although studies of equal ability in the hands may be out numbered by studies that relate hand preference with hand skill in visually guided conditions.

Studies in manual dexterity have for long tried to relate the preferred hand with actions of skill. When one hand is preferred over the other for various tasks, one can expect an enhancement in the ability of that hand over the other as is evident in the form of hand skills for tasks such as sewing or writing. These actions seem to indicate a relationship between hand preference and hand ability. Early attempts to know what is hand ability or skill in a certain action have emerged from intercorrelations of hand preference on a variety of tasks (Fleishman 1958; Fleishman & Ellison 1962). These when subjected to factor analysis have identified factors such as finger dexterity, manual dexterity, aiming movements or hand steadiness etc, thereby pronouncing the preferred or the right hand to be more skilled and faster (Barnsley & Rabinovitch 1970) than the left hand for most actions. A second approach examined the relation between speed and accuracy of aiming movements (Woodworth 1898). Though both hands were equally accurate at slower speeds, accuracy was greater for the preferred right hand at faster speeds both under visual control and eyes shut conditions.

Three theories have been proposed to account for the between hand performance differences in rapid aiming movements. One theory proposes that the preferred and non preferred hands differ mainly in their use of sensory feed back control. Flowers (1957) contrasted ballistic and control movements and found that the preferred and non preferred hands of strongly lateralized subjects achieved equal rates in a rhythmic tapping task though they differed with an increase in task difficulty. A second theory claims that the preferred hand is less noisy in its output function. Accordingly, increases in movement amplitude or decreases in movement time require a greater force, which leads to greater output variability and thus more errors. A third model which predicts a left hand advantage for larger target distances was supported by Todor and Doane (1978). The model expands on Flower's notion of feedback control by incorporating left hemisphere right hand superiority for sequential processing and right hemisphere left hand superiority for non adaptive parallel processing.

Subsequent explanations of handedness have indicated a relation between hand preference and hand ability (Annett et al. 1979; Peters 1980). They believed that if hand preferences can be coordinated with an independent measure of hand skill, then the understanding of handedness would be clarified. Both Annett et al.'s (1979) findings on the peg moving task and Peter's (1976, 1980) finger tapping tasks, independently reached the conclusion that the superiority of the preferred side is due to a more efficient control of the plan of action so as to effect a better motor output. In Annett's (1970a) association analysis study, hammering a nail was the action that was identified to be most discriminating between right and left preferences. Hammering a nail seems to require a consistency of hand use and efficiency, which in turn seemed to indicate skill. According to Annett (1970), moving a peg into its hole also required as much skill as hammering and therefore was a good indication of skill of the preferred hand. If the shapes of the distributions of hand preference and hand skill are considered, hand preferences show a J shaped distribution (Annett 1972) whereas the distribution of differences between the hands in skill is continuous and fairly normally distributed (Annett & Kilshaw 1983). Despite these differences in distribution, Annett (1970b) indicated that the difference between the hands for the peg moving times was highest for the right handers and decreased linearly for the mixed and left handers. Similar linear relationships between hand preference and skill were found among schoolchildren and undergraduates (Annett 1970a) among university students (Annett 1976) and among dyslexics and normal controls (Annett 1981). Besides Annett's studies, Peters and Durding (1978) also found a linear relationship between speed of finger tapping and hand preference, with the purely left handers being the slowest, and speed increasing orderly among nine degrees of hand preferences, with the purely right handers being the fastest group. Attempts to train the non preferred hand to equal the skill of the preferred hand have not been successful for peg moving (Annett, Hudson & Turner 1974) or for finger tapping (Peters 1976). Annett and Kilshaw (1983) found that degrees of mixed hand preferences between consistent right and consistent left are systematically related to degrees of L–R skill in the peg moving task, and Peters and Durding (1978) found a linear relationship between L–R mean tapping rates and hand preference. These findings led Annett (1985) to conclude that although practice can improve the performance of the non preferred hand, it does not alter the underlying natural asymmetry between the hands. A related notion to the above conclusion is that hand preferences are an outcome of eye- hand coordination and that eye- hand coordination is more efficient on the right than the left side of the body (Woodworth 1889; Annett et al. 1979; Peters 1976, 1980; Honda 1984).

A second group of studies do not consider handedness to be a unidimensional variable, but claim that hand actions may be controlled by groups of muscles that perform various actions and that the more skilled actions are more lateralized than less skilled actions such as picking up objects. For example, reviews of studies on the origin of handedness (Hopkins 1993) indicate that the earliest signs of hand preference appear to be task specific. Findings of reaching and grasping have shown the control of proximal muscles for reaching, or the distal segments of the hand for grasping. Adults tested with hand preference questionnaires (Beukelaar & Kroonenberg 1983) have indicated that hand tasks seem to cluster together according to the ways in which muscles and joints are involved in performing the task. Plato, Fox and Garruto (1984) found that hand preferences for tasks that were bimanual with whole body movement were discordant with hand movements in other categories. This indicates that there may be more than one dimension of hand preference. Healey, Lederman and Geshwind (1986) found certain behaviours that involve accuracy such as writing or ball throwing were exclusively performed by the right hand, whereas others that involved strength, such as lifting objects were less right biased. Peters (1990) observed that left handers who showed consistent left hand preferences had a strength advantage for the left hand and left handers who had inconsistent left hand preferences had a strength advantage for the right hand. However, Steenhuis and Bryden (1989, 1990) reported that hand preferences may distinguish themselves on the basis of skill such as handwriting, whereas less skilled tasks such as picking up objects are less lateralized. Ittyerah (1996) indicated that although during development hand preferences may group together into a single category of skill for each hand, the right hand may perform better for actions of accuracy such as writing or throwing (Healey et al. 1986) and the left hand may be more able for actions of strength (Healey et al. 1986; Peters 1990). The general ability between the hands may not differ, though the hands may differ according to task demands.

However, instances of hand advantage for tactual recognition have indicated that the preferred hand may not always be the skilled hand. For example, Witelson (1974) found that boys from grades 1 to 8 showed a left hand advantage for forms and a right hand advantage for pairs of letters they had felt similarly. Oscar-Berman et al. (1978) reported right hand superiority for letter identification, no difference for digit identification and a left hand effect for line orientation. Furthermore, a left hand advantage for various forms of non verbal spatial tasks (Kumar 1977; Dodds 1978) seemed to show right hemisphere superiority for tactual recognition. When the right hand performs better, the conservative conclusion is that some verbal element of the task is processed since language is mainly controlled by the contralateral left hemisphere. However not all tactual performance depends on systems which are entirely lateralized as in instances of discrimination of roughness, texture and passive tactual discrimination (Gazzaniga & Ledoux 1978). Besides there is evidence that the tactile matching of height, depth, breadth and volume of three dimensional objects does not differ between the hands and hand preferences do not dictate hand ability (Ittyerah 1993) in blind or sighted blindfolded conditions over successive periods of development.

Other studies revealed that a particular hand may be skilled for certain activities. Tan (1990) found that both hands of left handers had equal capacities for visuo-motor learning. Peters (1991) indicated some evidence of left handedness and mathematical giftedness or strong right handedness and lack of mathematical giftedness. Gotestam (1990) found a high frequency of left handedness among architecture students. Natsoupoulos, Kiosseoglou and Xeromoritou (1992) supported Geshwind and Galaburdas (1987) postulate that left handers are superior to right handers in spatial ability and Annett and Manning's(1990) findings that right hand preference skills have costs for intellectual ability. DeLuca, Burright and Donnvick (1990) found in the concurrent performance of a verbal and spatial task that the spatial task was performed more often by the left hand and the verbal task was performed more often by the right hand. Gordon and Kravetz (1991) indicated that high performing right handed males and left handed females were better on a visuo-spatial task, while left handed males and right handed females performed better on verbo- sequential tasks. The left hand was also found to respond to emotional situations (Hatta & Koika 1991) as in instances of left handed picking up of babies by mother monkeys showing that the left hand opposite the right hemisphere is dominant for emotional specialization. Salk (1960) was the first to report from casual observation of a mother rhesus monkey that this female held her new born on the left side 40 times out of a total of 42 observations, often with the newborn's ear pressed against her heart. This cradling bias was also evident among human mothers. Salk (1960) found a left side cradling bias at a population level in 80% of human mother-infant pairs, regardless of the mother's handedness. This left side bias was spontaneous on the part of the mother. Salk proposed that the heart beat of the mother being louder on the left side may be a comforting stimulus to which the infant may be imprinted. Following Salk's pioneer findings, others have reported the same left ward maternal cradling in human populations (Saling & Kaplan-Solms 1989).

The question as to whether a particular hand is more skilled than another has not been satisfactorily answered. In nonprehensile tasks such as Braille reading, type writing or piano playing or for prehensile actions of juggling, the hands have a complementary role in task performance. This indicates that the skill is not lateralized, but rather, that task requirements dictate hand actions. For example, there was some initial confusion as to whether Braille was predominantly read by one hand. Superior Braille performance was reported for the left hand (Hermelin & O'Connor 1971; Rudel, Denckla & Hirsch 1977), at other times for the right hand (Fertsch 1947), or for neither hand (Bradshaw, Nettleton & Spehr 1982; Millar 1977) and for two handed reading (Foulke 1982). Millar (1987a) argued that in so far as reading levels were reported, the discrepant findings indicate a pattern that conforms to the notion that highly proficient reading depends mainly on verbal strategies and skill (right hand/left hemisphere advantage); less proficient reading demands attention to spatial coding of the physical characters (left hand/right hemisphere advantage), while early in learning subjects rely on dot density or texture features of Braille characters. Millar (1984) found that hand advantages depend on a combination of task demands and individual strategy preferences associated with individual proficiency or retardation. Furthermore, for tasks that demand lateral movements, two handed reading was superior to reading by either hand, irrespective of absolute or relative reading proficiency. Reading Braille involves a number of subsidiary spatial and verbal skills, and this would presumably involve processing by both hemispheres (Gazzaniga & Ledoux 1978). The question as to whether the two hands process text simultaneously was examined by Millar (1987). It was found that the two hands rarely do the same thing at the same time. The findings suggested that the advantage for fluency depended on using both hands flexibly for all subsidiary verbal and spatial skills needed in reading Braille prose texts. Touching a letter by one hand was paired most often with simultaneous touching of a space between letters or between words by the other hand. Simultaneous touching of the letters by both hands was least frequent and did not differ from simultaneous touching of gaps between letters. Fluency does not depend on simultaneous processing of the same type of information, but in fast intermittent alternation in function between the two hands. Such a complementary relationship between the hands can also be found in other actions such as piano playing, type writing and juggling, all of which indicate that general lateralization does not affect ability (Ittyerah 1993).

5.3 Hand ability

Bimanual coordination is an instance in which the hands function together and indicate the complementary use of both hands. However, attempts to explain the outcome of bimanual coordination have usually referred to findings of an asymmetry between the hands, so as to reveal a dominant and a subordinate hand relationship. The conventional wisdom is that because the cortical hemispheres are asymmetrical, though more in function than in structure, these effects of lateralization must also be manifest in hand actions or related behaviours. Thus studies in bimanual skills usually report that though the preferred hand is more skilled, the outcome of bimanual coordination will also depend on the task, and competition and cooperation between the hands to maintain an even performance. The finding that general lateralization does not affect ability (Ittyerah 1993, 2000, 2009) indicates, that although one may have a hand preference, there is an equipotentiality between the hands. In nonprehensile tasks such as Braille reading, Millar (1987) found that fluent Braillists use both hands in intermittent alternation for processing text. As to whether this is also true for prehensile actions can be known by testing for hand ability.

In an earlier study (Ittyerah 1993) the development of hand preference and hand ability in congenitally blind children between the ages of 6 and 14 years was compared with that of sighted blindfolded children at comparable ages so that any effects of laterality could be observed at different stages in development. It was expected that the hand preferences of the blind children would not be very different from that of the sighted, as the role of vision in determining hand preferences has not been particularly well demonstrated. For a tactile task with three dimensional objects, the two groups may not differ in their hand abilities for detecting spatial differences in height, length, depth or volume. If a right hand advantage for the three dimensional task could be shown, it may implicate culture and training, whereas a left hand advantage may be implicative of a corresponding tendency for lateralization (Gazzaniga & Ledoux 1978) reflecting an underlying right hemisphere specialization for spatial discrimination. Annett (1985), for example, has argued that there is an underlying natural asymmetry that does not alter with training the non preferred left hand. Left hand advantages for spatial material have been reported for example by Witelson (1976) for normal boys in dichaptic shape recognition and by Nebes (1972) for right handed commissurotomy patients in a form perception task. The study tested whether similar effects occur with both blind and sighted children. Results of hand preferences (Figures 5.1 and 5.2) revealed that the left hand preferences of both blind and sighted blindfolded children show the typical J curve with respect to the distribution of preference across numbers of tasks, as Annett (1972) observed in her studies.

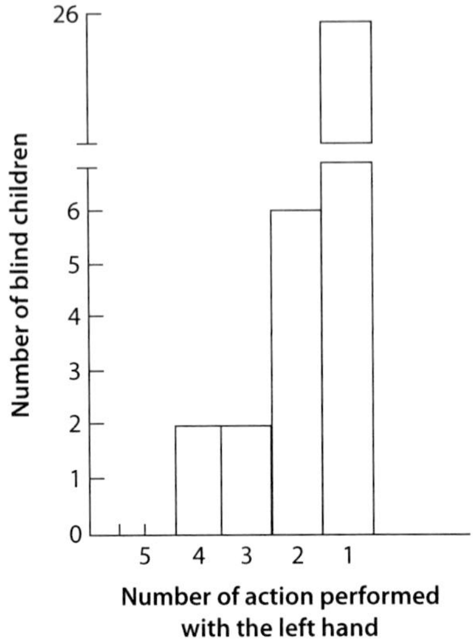

Figure 5.1. Numbers of left hand preferences of the blind children for performance tasks 1–5

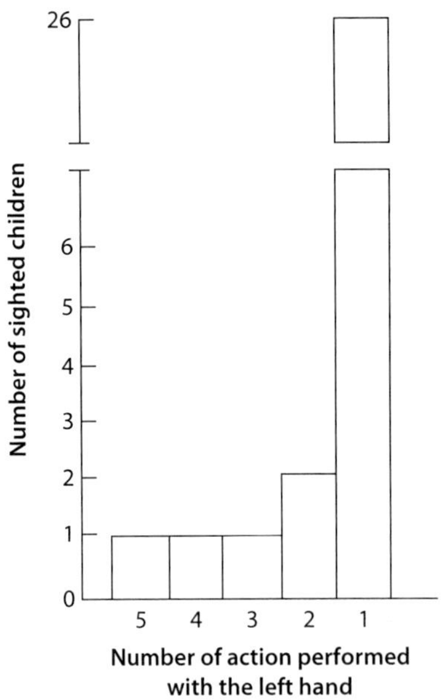

Figure 5.2. Numbers of left hand preferences of the sighted children for performance tasks 1–5

From Ittyerah, M. (1993). Hand preferences and hand ability in congenitally blind children. Quarterly Journal of Experimental Psychology, 46 (1), 35–50. Copyright © The Experimental Psychological Society. Reprinted by permission of (Taylor and Francis Ltd, http.//www.tand f online.com) on behalf of the Experimental Psychological Society

Though no blind child performed all five tasks with the left hand, the number of children performing four or fewer tasks with the left hand increased more gradually in the blind. Verbal reports of left hand preferences of tasks performed in daily life, such as combing hair and brushing teeth were less than half of those of the performance tasks. A comparison of the percentages of the left hand preferences of the two groups indicated that blind children performed nearly 2% more of the tasks with their left hands.

The hand preferences were converted to a laterality index (LI) (McManus et al. 1988) derived from the performance scores for each child. Scores for direction and degree of lateralization were calculated for each child. The scores range from −1 through 0 to +1 and those scoring greater than 1 were taken as right handers. The degree of lateralization was considered to be the absolute value of LI (range 0 to 1). The distributions of hand preferences (Figures 5.3 and 5.4) for blind and sighted children are not very different, suggesting that vision is not a necessary condition for determining hand preference.

Besides the choice of the preferred hand, the tactile perception of three dimensional stimuli indicated differential hand skills in discriminations of length, breadth, height and volume of an object. Detection of spatial qualities in an object has been found to be superior with the left hand (Nebes 1972) reflecting superior right hemisphere performance (hand abilities being controlled by the contralateral hemisphere; Brinkman & Kuypers 1973). Thus it should be expected that the matching of bricks on the basis of their spatial dimensions would be more accurate with the left hand. The question is whether (a) hand ability for spatial discrimination is related to the extent to which the same hand is used

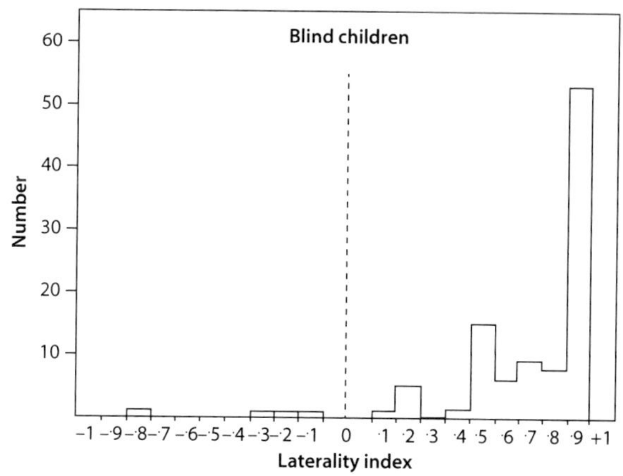

Figure 5.3. Bimodal distribution of laterality scores calculated from the 5-item performance battery for blind children

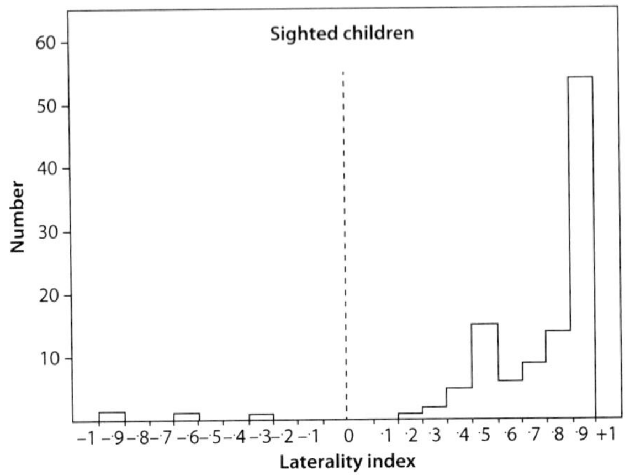

Figure 5.4. Bimodal distribution of laterality scores calculated from the 5-item performance battery for slighted children

From Ittyerah, M. (1993). Hand preferences and hand ability in congenitally blind children. Quarterly Journal of Experimental Psychology, 46 (1), 35–50 Copyright © The Experimental Psychological Society. Reprinted by permission of (Taylor and Francis Ltd, http.//www.tand f online.com) on behalf of the Experimental Psychological Society

for a manual task and thus a phenomena 'sui generis' (McManus et al. 1988) or (b) because the detection of spatial qualities is primarily controlled by the right hemisphere, this would result in left hand discriminations being more accurate than those of the right hand.

The tactile perception of three dimensions was tested with a set of eight wooden bricks that varied in height, length, breadth and volume. The bricks were presented in pairs to each child. The children were tested individually and the instructions were read aloud. The instructions encouraged the child to feel each brick with one hand at a time. Following this, another brick was placed on the table before the child. The child was required to feel the brick with the same hand and to say whether the second brick was the same or different from the previous brick. If the response was different the child was required to indicate if the difference was in height, breadth, depth or volume.

The performance scores showed that height, depth and breadth judgments all increased in accuracy with age, whereas volume judgement seemed hardly to change. (However, these effects were not found to be significant). The results indicated that the blind and sighted children did not differ in the tactile perception of three dimensional stimuli. The results suggested that accuracy of tactile matching is strongly related to age; there is no effect of blindness in this ability, nor is there an effect of the hand used in this task. Although the perception of spatial extent was not affected by the hand of responses, each hand showed an effect of order on accuracy. The left hand was more accurate in matching singular dimensions such as height followed by depth, volume and breadth. In contrast, the right hand was more sensitive to the perception of the entire object in space (that is, to its volume), followed by singular dimensions of height, depth and breadth. These hand differences for order of perception suggest that spatial judgements are performed differently when using each hand.

In order to test whether degree of lateraliation related to subjects' ability on the task, computations were made of the number of responses made with the right and left hands, the total correct responses (R+L) and the difference between the right and left responses (R–L) for each subject. These were then correlated for all the subjects with age, blindness and absolute laterality scores (LI). There was no correlation of the R–L scores on the tactile perception scores with the laterality score but there was an association between overall performance accuracy (R+L) and the degree of lateralization. This association reflected the fact that both the degree of lateralization and the overall ability of the subjects were correlated with age.

The results of the hand preference tasks failed to show any difference between the blind and sighted children in the direction of their hand preference. Visual impairment does not therefore affect the choice of the preferred hand in children. The degree of handedness increased with age and appeared to be generally linear, although the blind children were somewhat less lateralized than the sighted at most ages under ten years. Lateralization tends to increase with age in childhood (McManus et al. 1988). There is no evidence from this study that the degree of lateralization in blind or in sighted children is necessarily a function of learned experience with an asymmetric world. It has been suggested that the extent to which handwriting is used in childhood (in schooled, sighted children) may predict the degree of lateralization (Batheja & Mcmanus 1985). Also the notion of eye- hand coordination (Peters 1983; Honda 1982) in developing hand preferences was not evident. If that were the case, then the onset of any visual impairment should affect the direction or degree of hand preference. The absence of group differences in the brick matching task showed that blind children were not more sensitive than sighted children when perceiving tactile differences of height, depth breadth or volume of objects. Lack of sight does not necessarily improve touch perception.

A subsequent study (Ittyerah 2000) tested the relationship between hand preference and hand ability during development in blind and sighted conditions, to learn about the role of vision in hand preference and ability in temporary blindfolded conditions. The hand preference test consisted of twenty performance tasks that varied in hand use from unimanual to bimanual preferences. Most of the hand preference tasks were selected according to Plato et al.'s (1984) categories and Geshwind's suggestions in Healey et al. (1986). There were a total of 20 handedness items and each child was required to perform each task three times at random. Thus there were a total of 60 hand actions for each child.

The hand preferences of the children revealed that the blind had more right hand preferences (98%) as compared to the sighted (91%) though these differences were not significantly different. There were no differences in the hand preferences of the boys and girls in that both had more right hand preferences. Similarly neither the hand preferences of the blind and sighted boys nor the blind and sighted girls differed significantly. The only effect was that of age showing lateralization increased with age in development. Therefore the groups do not differ in the direction of their hand preference.

To test for hand ability, three hand ability tasks, a sorting task, a finger dexterity task and the Minnesota rate of manipulation test were presented to each of the children.

Sorting task

The sorting test consisted of four sets of items, ten playing cards, ten plastic tokens, ten paper clips and ten buttons. Every child (blind and sighted blindfolded) was required to sort the forty items into four categories with the left and right hands separately. The time taken to sort the items was recorded for each hand. This tested the ability of the child to discriminate shapes by touch and arrange them into groups.

Finger dexterity test

This tested the precision of coordination between the thumb and forefinger for manual dexterity. It consisted of a wooden board in which there were 40 holes arranged in five rows and eight columns, half an inch apart. At the top right corner of the board there was a small tray with pins one inch long and 1/16 inch in diameter. Each child was required to pick a pin and insert it into a hole with the left and the right hands separately. This test helps to predict success and efficiency in tasks demanding manual dexterity and fine precision finger movements.

Minnesota rate of manipulation test

This test was adapted for blind participants by Roberts and Bauman (1994) and involves displacing and turning pegs into holes. It consists of a box with pegs arranged in its hole along the length and breadth in rows and columns. Every child was required to turn each peg, lift it from its hole and replace it with one hand. The children were required to perform this manipulation with their left and right hands and the time taken to replace all the pegs correctly with each hand was recorded. The test provides an assessment of the child's manipulative skill and dexterity for a range of hand actions. The test is effective in a range of tasks (Roberts & Bauman) requiring speed and accuracy in the coordination of arm and hand movements.

The results indicated no differences between the hands, although there were some group differences. As the tasks differed from one another in the assessment of manual dexterity, each task was analysed separately. There were no left and right hand performance differences for any of the tasks ($p > .01$), though in the finger dexterity and the Minnesota rate of manipulation, the sighted blindfolded group performed better than the blind children ($p < .01$). To find the relation between hand preference and ability on the task, correlations computed between the LI scores of the children and their performance on each task with each hand indicated that only age was significant ($p < .05$). Thus hand preferences did not affect ability on the task.

Blind children improve in proficiency when they are allowed sufficient practice with spatial tasks (Millar 1994), indicating that task performance cannot be solely attributed to vision. However, whether practice will show gains only from the preferred hand remains to be established. In tasks such as type writing, piano playing or Braille reading (Millar 1987) the hands perform as well as each other. These are however non prehensile movements where the object is manipulated by the hand or the fingers and not grasped in the hand. In tasks using prehensile movements when the object is partly or wholly held, the evidence is sparse. For example, it has been found in blind conditions, that neither the congenitally blind nor the sighted blindfold children differ with their hands in the tactile perception of height, length, depth or volume of bricks nor in hand skills that assess different types of dexterity such as sorting and placing objects and therefore the hands do not differ in ability (Ittyerah 1993, 2000). However, since the effects of practice with the hands is as yet inconclusive at least for tasks using prehensile movements, the question as to whether practice will show gains only for the preferred hand remains to be sought. The question whether eye hand coordination is a prerequisite in performing these tasks after periods of practice was of interest as well.

It was therefore hypothesized that if the preferred hand gains from practice in the sighted blindfolded children, skill might possibly depend on visuo-motor control (Annett et al. 1979; Jeannerod 1997). If the preferred hand gains from practice in the blind, then visuo-spatial information is not necessary for improved haptic performance and children may well gain in proficiency (Millar 1994). If the preferred and non preferred hands do not differ with gain (Ittyerah 1993, 2000) after practice in either the blind or the sighted blindfolded children, then any asymmetry between the hands (Annett 1985) is not evident, and the blind children may perform as well as the sighted children and gain in proficiency in the haptic processing of spatial information (Millar 1994). Consequently, successful performance with the left and right hands may also predict vocational success in tasks requiring hand skill.

Blind and sighted blindfolded children in three age groups (6–8, 10–11 13–15) each completed a personal information page containing age, gender, class, onset of blindness, degree of blindness, mental and physical condition, hand preference, handedness of the parents and siblings, and occupation and income of the parents. All the children belonged to 'middle class' homes (a family income of Rs.10, 000 to Rs.20, 000 per month). The experimenter assisted those children, particularly the very young that were not able to complete the data sheet with the help of the teacher. Each child's handedness was assessed by a ten item test of hand preference. This was followed by four tests of hand skill.

The handedness task was similar to that described in Ittyerah (2000) and contained items that assessed the repertoire of hand skills in daily life. There were four hand ability tasks, a sorting task, stacking task, finger dexterity and the Minnesota rate of manipulation test. The tests required an object to be held partly or wholly within the hand. The time taken to perform each task for every child was recorded in seconds with a stop watch.

Sorting task

This task was designed to measure each hand's ability to tactually discriminate size, shape and texture of objects and arrange them in groups on a horizontal plane. It consisted of 4 sets of items: 10 small plastic bottles (2 inch length), 10 curtain rings (4 inch diameter), 10 buttons (1 inch diameter) and 10 small batteries. These items were mixed on a tray and presented to each child. Every child (blind or sighted blindfolded) was required to sort all the 40 items into four categories separately with her/his left and right hands. While the child performed with one hand, the other hand remained at the side.

Stacking task

This tested the ability of each hand to lift and place objects on top of each other in a vertical plane. It consisted of six light cardboard boxes (9 cm × 7 cm × 5 cm). The child was required to pile or stack them one upon the other with each hand. Successful completion of the task indicated the speed and accuracy of the hand for lifting and placing objects on top of each other. The task required the child to anticipate the amount of force required to lift each box and place it upon another without dislodging the stack on which it was to be placed. It was of interest to know whether eye hand coordination is a necessary condition to perform this task.

The finger dexterity test and the Minnesota rate of manipulation test were the same as those described in Ittyerah (2000).

Each child was required to sit at a table in a quiet room in the school and instructed separately for every task. The instructions required the child to perform each task as described in the previous task performance section, separately with the left and right hands and the performance times were recorded in seconds. The order of the tasks and hands were counterbalanced across groups and ages of the children. This was followed by a practice period of four months for each group during which time each child was required to practice each task three times with each hand in front of the experimenter. The performance times were not recorded during the practice trials. Three practice trials for each hand and task ($3 \times 4 = 12$) were considered to produce performance effects above chance. The children were post tested and the performance times of each task and hand were recorded in seconds. The sighted blindfolded children practiced the tasks with their blindfolds. The order of the two groups of children tested was the same in pretest as in the posttest.

The results have been analysed separately for the hand preference and the performance tasks. The analysis of hand preference indicated that there were more lefthanders in the sighted group. In fact there were no left handers in this group of blind children. Although no blind child performed all the actions with the left hand, 7 of the 90 children performed at least 2 actions of a possible 30 with the left hand.

The hand preferences of the children revealed that the blind children had more right hand preferences (100%) as compared to the blindfolded sighted (93%). Chi square test used to compute the direction of hand preference indicated that this difference was not significant ($\chi^2 = 84.41$ (df 1,269) $p > 0.05$). There are no differences between the boys and girls in that both have more right hand preferences ($\chi^2 = 81.69$ (df 1, 269) $p > 0.05$). To explore the effect of age, the children were divided into three age levels, 5–7 years, 9–11 years and 13–15 years and χ^2 (Chi Square) revealed that age was significant ($\chi^2 = 21678.25$ (df 2,269) $p < .01$), indicating more right hand preferences with age. Therefore the groups do not differ in the direction of their hand preference.

Further, there were no differences between the degree of lateralization of the blind (mean = 97.8) and sighted blindfolded groups (mean = 97.8), ($F(1, 144) = .0001$; $p > 0.05$). The effect of age was significant ($F(8, 144) = 4.338$; $p < .001$) showing that lateralization increases with age in development (mean age groups 5–7 = 93.72; 9–11 = 99.6; 13–15 = 100). The main effect of gender was not significant ($F(1,144) = .54$; $p > 0.05$). None of the interactions were significant.

Outcome of practice

Performance differences for all the tasks were computed separately for each group and hand by assessing the difference between the pre- test times and the post-test times. Performance differences were calculated as percentage change for each child in the following manner:

$$\% \text{ change} = ((\text{post-test} - \text{pre-test})/\text{pre-test}) \times 100$$

Since there were 180 children (90 in each group), the post-test differences were calculated separately for each child's left and right hands for each of the four tasks. In all there were 180 × 4 = 720 post-test scores for the left hand and 180 × 4 = 720 post-test scores for the right hand. The means of the percentage change revealed gains for the blind children.

Since the tasks varied with difficulty and each tested for different types of dexterity, each task was analyzed separately. Fleischman (1958) for example has indicated that measures of manual proficiency correlate poorly and there are different dimensions of unimanual proficiency such as strength, speed, accuracy and precision.

Sorting task

The means and standard deviations of the percentage change with practice in the sorting task for the left and right hands of the blind and sighted blindfolded children are presented in Table 5.1. Table 5.1 indicates that the blind children gained by 25% and the sighted blindfolded children lost by 52%.

Table 5.1. Mean and SD for the sorting task for group, hands and ages

	Blind children				Sighted blindfolded children			
Age	Mean LH	S D LH	Mean RH	S D RH	Mean LH	S D LH	Mean RH	S D RH
5	−41.4	8	−45.1	7.1	1.9	31.8	5.1	31.5
6	−39.7	4.2	−39.3	8.9	15.8	29.8	89.7	70
7	−19.6	12.1	−21.5	7.7	95	62.9	30.4	39.2
9	−35.4	12.6	−23.4	13	31.7	38.2	29	60.6
10	52.8	19.3	−38.5	9.4	95.2	92.8	34.5	23.7
11	−33.6	12.3	−38.9	13.6	56.5	39.8	53.1	52.1
13	−13.7	14.5	−21	26.9	55.5	33.8	44.3	59.4
14	−36	10.2	−19.8	22.9	80.5	104.8	70.9	58.2
15	−18.5	18.2	−12.7	27.7	68.2	39	78.6	79.3
Mean	−20.6	12.4	−28.9	12.7	55.6	52.5	48.3	52.5

From Ittyerah, M. (2009) Hand ability and practice in congenitally blind children. Journal of Development and Physical Disabilities, 21, 329–344.

A mixed factor ANOVA (group (2) by age (9) by hand (2)) with repeated measures on the last factor with the times of the percentage change for the sorting task indicated that the blind group (25%) gained more than the blindfolded sighted group (52%) following practice (F (1, 162) = 307.9; p < 0.001). The main effect of age was significant (F (8,162) = 6.58; p < 0.001) showing more performance gain at the younger ages. Post hoc Duncan's multiple range test for means indicated that ages 5, 6, and 7 gained more than the children at the older ages (p < .05). The main effect of hand was not significant (F (1, 162) = 3.1; p > 0.08), (left hand = 17.52%; right hand = 9.7%) indicating no significant differences between the hands after practice. The group by age interaction was significant (F (8, 162) = 2.17; p < .03). Post hoc Newman Keuls test revealed that the blind children at ages 5 and 6 gained more than the older children in both groups (p < .05).

Stacking task

The means and standard deviations of the percentage change with practice in the stacking task for the left and right hands of the blind and sighted blindfolded children are presented in Table 5.2. Table 5.2 indicates that the blind children gained by 17.3% and the sighted blindfolded children lost by 12%.

A group (2) by age (9) by hand (2) ANOVA with repeated measures on the last factor with times of the percentage change following practice in the stacking

Table 5.2. Mean and SD for the stacking task for group, hands and ages

	Blind children				Sighted blindfolded children			
Age	Mean LH	S D LH	Mean RH	S D RH	Mean LH	S D LH	Mean RH	S D RH
5	−28.3	75.4	−26.7	50.9	44.5	22.9	−65.3	10.5
6	−34.9	25.3	−6.1	50.7	5.9	39.8	−9.8	32.8
7	.86	70	−17.2	52.6	88.4	19.3	81.3	37.6
9	−13.5	49.3	−8.9	57	47.8	30.4	−4.7	25.4
10	2.8	65.8	14	67.2	5.4	32.7	5.2	45
11	−7.5	73.9	6.7	64.3	−8.7	41.8	−26.5	33.4
13	−60.7	19.4	−30.8	29.3	34.4	64.3	23.6	70.2
14	−33.1	49.3	−23.1	65.5	−1.1	3.5	−10.7	44.4
15	−12.3	49	−32.3	23	43.5	52.6	32.7	13
Mean	−20.7	53	−13.8	51.2	28.9	34.1	−3.39	34.7

From Ittyerah, M. (2009) Hand ability and practice in congenitally blind children. Journal of Development and Physical Disabilities, 21, 329–344.

task indicated that the blind group (–17.3%) gained more than the blindfolded sighted group (12%), (F (1, 162) = 48.8; p < 0.001). The main effect of age was significant (F (8, 162) = 6.2; p < 0.001) showing more gain for the youngest group of children. Post hoc Duncan's test indicated that children at ages 5, 6, and 7 gained more than those at the older ages. The main effect of hand was not significant (F (1, 162) = 3.4; p > .07) (left hand = 4.07%; right hand = –5.5%). The group by age interaction was significant (F (8. 162) = 6.5; p < 0.001). Post hoc tests (Newman Keuls) indicated that the blind children gained more than the blindfolded sighted group at most ages and the blindfolded sighted group gained the least at ages 7, 9, 13, and 15 (p < .05). The group by hand interaction was significant (F (1, 162) = 10; p < 0.002) showing more gains for the blind group for the left and right hands p < .05).

Finger dexterity task

Means and standard deviations of the percentage change with practice in the finger dexterity task for the left and right hands of the blind and sighted blindfolded children are presented in Table 5.3. Table 5.3 indicates that the blind children gained by 14.6% and the sighted blindfolded children lost by 52%.

A group (2) by age (9) by hand (2) ANOVA with repeated measures on the last factor, for the finger dexterity test indicated that the blind children gained

Table 5.3. Mean and SD for the finger dexterity task for group, hands and ages

	Blind children				Sighted blindfolded children			
Age	Mean LH	S D LH	Mean RH	S D RH	Mean LH	S D LH	Mean RH	S D RH
5	–15	9.8	–10	6.1	31.4	6.5	26.1	1.3
6	–9	18.9	–15	17.1	26	3.2	22.5	2.3
7	–7.3	15.8	–10.3	11	64	3.1	61	4.6
9	–16.5	24.3	–21.6	13.7	31.6	7.3	27.3	3.7
10	–26.2	20.5	–22.1	14.4	31.4	1.8	32.1	3.1
11	–7.4	20.5	–8.9	13.6	21.4	4.8	14.7	4.1
13	–17.5	24.2	–20.5	18.2	96.4	9	94.2	11.3
14	–15.9	29.7	–17.8	13	84.9	7.4	79.8	8
15	–13.8	–14.6	–8	5.8	90.3	3.8	101.4	7.5
Mean	–14.3	19.8	–14.6	12.5	53	5.2	51	5.1

From Ittyerah, M. (2009) Hand ability and practice in congenitally blind children. Journal of Development and Physical Disabilities, 21, 329–344.

from practice (–14.6%) over the sighted blindfolded children (52.1%) ($F(1, 162) = 1400.2$; $p < .001$). The main effect of age was significant ($F(8, 162) = 36.2$; $p < .001$) showing more practice gain at the younger ages. Post hoc Duncan's test showed that children at ages 5, 6, and 9 gained more than those at the older ages ($p < .05$). The main effect of hand was not significant (F (1, 162 = 3.3; $p > 0.07$) (mean left hand = 19.4%; mean right hand = 18.1%) showing no differences between the hands in percentage gain for finger dexterity. The group by age interaction was significant ($F(8, 162) = 37.7$; $p < 0.001$). Post hoc test revealed that the blind children at ages 5, 6, 7, 9, 10, and 13 years gained more than those at other ages ($p < .05$). The age by hand interaction was significant ($F(8, 162) = 4$; $p < .0003$). Post hoc tests indicated that the right hands of the children at ages 6, 9 and 11 gained more after practice.

Manipulation task

Means and standard deviations of the percentage change with practice in the manipulation task for the left and right hands of the blind and sighted blindfolded children are presented in Table 5.4. Table 5.4 indicates that the blind children gained by 8.8% and the sighted blindfolded children lost by 15.6%.

Furthermore, a group (2) by age (9) by hand (2) ANOVA with repeated measures on the last factor for the manipulation task indicated that the main

Table 5.4. Mean and SD for the Minnesota rate of manipulation task for group, hands and ages

	Blind children				Sighted blindfolded children			
Age	Mean LH	S D LH	Mean RH	S D RH	Mean LH	S D LH	Mean RH	S D RH
5	8.8	26.5	–25.4	5.9	13.5	22.4	–3.3	19
6	–8.5	19.6	–12.1	29.4	1.1	18.4	1.03	18.2
7	2.5	11.6	–12.1	22.1	–3.6	10	3.3	40
9	–15	11.5	–16.2	16.3	4.7	17.4	13.2	57.1
10	.89	16.5	–4.7	29.7	–8.2	26.3	–1.8	36
11	–5.3	11	–14.4	18.3	–.4	19	.4	47.3
13	–5.7	8.4	–10.4	23.1	17.7	23.7	24	35.6
14	–10.5	11.8	–10.1	18.9	10.8	15.2	36.5	73.4
15	–9.5	15.7	10.6	22.7	70.6	45.1	101.1	65.9
Mean	–4.7	14.7	–10.5	20.7	11.8	22	19.4	43.6

From Ittyerah, M. (2009) Hand ability and practice in congenitally blind children. Journal of Development and Physical Disabilities, 21, 329–344.

effect of group was significant (blind group = –8.8%; sighted group = 15.6%), ($F(1, 162) = 44.7$; $p < .001$). There are performance gains with age ($F(8, 162) = 6.2$; $p < .001$), though there are no differences between the hands after practice (left hand = 3.6%. right hand = 3.3%) ($F(1, 162) = .02$; $p > 0.9$). Post hoc Duncan's test showed that children across ages 5 and 11 gained consistently more than those at older ages ($p < .05$). The group by age interaction was significant ($F(8, 162) = 7.2$; $p < 0.001$). Post hoc test revealed that the blind gained more than the sighted blindfolded at all ages ($p < .05$). The age by hand interaction was significant ($F(8, 162) = 2.4$; $p < .02$) Newman Keul's post hoc test revealed that the right hand of the children gained more than the left hand at ages 6, 7, 9, 10, and 11($p < .05$). The group by hand interaction was significant ($F(1, 162) = 10$; $p < .002$). Post hoc tests revealed that the right and left hands of the blind children gained more than that of the sighted blindfolded children ($p < .05$). The results on the whole indicated that though there may be performance differences between the groups after practice with increasing effects with age, the performance of the left and right hands do not differ with practice in any of the tasks for either the congenitally blind or the blindfolded sighted children.

In order to test whether the degree of lateralization related to subject's ability on the task, correlations were computed between (1) age and LI (laterality index) scores (.31; $p < .05$), (2) total scores (left and right hand) and LI (.02), (3) right hand sort and LI (.12), (4) left hand sort and LI (.14), (5) right hand stack and LI (–.13), (6) left hand stack and LI (–.11), (7) right hand dexterity and LI (.05), (8) left hand dexterity and LI (.02), (9) right hand manipulation and LI (.07), (10) left hand manipulation and LI (.07). The correlations indicated that only the effect of age was significant. Thus there was no effect of hand on ability in either blind or sighted conditions.

The results of the study indicated a gain in performance times for the blind children for all the tasks with their right and left hands and no significant differences between the hands for any group. Percentage gains were more for the blind than the sighted blindfolded children, indicating that practice improves skill in the absence of vision (Millar & Ittyerah 1991). The sighted blindfolded had more losses than gains in the post-test, probably because the blindfolds hampered performance, indicating their dependence on vision in interacting with the environment.

The finding that the congenitally blind children gained from practice indicates that performance is not solely confined to benefits from vision. The use of self- referential coding is evident in the blind and to a lesser extent in the sighted blindfolded children. It also indicates that the losses of the sighted blindfolded children may have been caused by the difficulty of having to rely on body- centred self- referent codes that are usually ignored in their sighted existence. Millar (1981) for example found that reliance on self referent coding was greater in the congenitally blind children in the recall of rotations than among their sighted blindfolded cohorts and the type of error that showed self referent coding related specifically to prior visual experience and not to age. This indicates that long term haptic experience tends to elicit more self referent coding than prior visual experience. The sighted blindfolded children incurred more losses than gains with practice because of the difficulty of performing the tasks in the absence of vision. Furthermore, since most of them were right handed, performance with the left hand may have been relatively difficult in the absence of vision, although the hands do not differ in performance with practice in blind or blindfolded conditions. Therefore lateralization or a consistent hand preference is different from hand ability.

Blind children have been found to benefit from practice and gain on the Revised Peabody Mobility Scale (Harley, Wood & Merbler 1980) and in tests of cognitive development (Lebron-Rodreguez & Pasnak 1977). This indicates gains in proficiency in the absence of vision. Thus the difference between the blind and sighted children may be in their strategies for attaining information and is not a function of visuo-motor control (Annett et al. 1979; Peters 1980; von Hofsten 1982; Jeannerod 1997). The role of haptic and visual experience may also depend on aspects of the task. Postma, Zuidhoek, Noordzi and Kappers (2007) examined the roles of visual and haptic experience in different aspects of haptic processing of objects in peripersonal space. Early blind, late blind and sighted blindfolded subjects had to haptically match shapes to the cutouts in a board as fast as possible. Both blind groups were faster than the sighted blindfolded group. The blind were superior to the others for speeded matching after rotation of the stimulus frame indicating that the role of haptic and visual experience appears to depend on aspects of the task. Congenital blindness is a rare human model to explore the role of experience following early sensory deprivation. Though the spatial attention abilities of the congenitally blind were similar to that of the sighted controls for auditory and tactile stimuli, the blind had shorter reaction times than the sighted for the detection of spatial targets in both modalities for both selective and divided attention (Collington & De Volder 2009). The blind were better in both attention tasks, and this informs and broadens our knowledge of the mechanisms underlying cross-modal compensation.

Although complete absence of sight does reduce the total amount of available information, the main information that is missing is quite specific. Body- centred reference cues are available as much to sighted people as to blind people, though the advantage of sighted conditions is that external and body centred inputs coincide. Studies in which blind and sighted children were required to match tactile stimuli with the left and right hands separately have indicated that the hands do not differ in ability for matching the length, breadth, height and volume of three- dimensional bricks of varying sizes (Ittyerah 1993) and can perform different manual dexterity tasks such as sorting, finger dexterity and the Minnesota rate of manipulation test (Ittyerah 2000). Although there were differences between the blind and sighted children in task performance, the left and right hands of the blind or sighted children did not differ in speed or accuracy. Results indicated that performance improved with age, though the hands did not differ. A follow up study conducted to know if practice would facilitate the performance of the preferred hand indicated an increasing post- test gain with age for the groups on the tasks, though the hands did not differ in performance either before or after practice (Ittyerah 2009). These consistent findings indicate that even if there is a hand preference (Ittyerah 1993, 1996, 2000, 2009), the general ability of the hands in most tactile tasks does not differ. Therefore hand preferences are not related to hand ability in prehensile tasks as well. Some earlier evidence of comparable performance with the left and right hands was found for curvature discrimination (Kappers & Koenderinke 1996) in right handed subjects. The authors tentatively concluded that the curvature representations of either hand are not independent of each other. Regardless of the conclusion, systematic data indicate no significant performance differences between the hands, thus lending support to the present theoretical notion of equipotentiality between the hands. Furthermore, lack of sight does not affect hand ability, just as vision neither determines the direction nor the degree of hand preference (Ityerah 1993, 2000, 2009).

Speculations on the effects of practice can also suggest different possibilities that practice would facilitate the skill of the preferred hand and skill might possibly depend on visuomotor control (e.g. Annet, Golby & Kay 1958; Jeannerod 1984, 1994, 1997a, 1997b) or that visuospatial information is not necessary for improved haptic performance and subjects may well rely upon self referent cues (Millar 1994) and gain in proficiency. However if the preferred and non preferred hands do not differ in visually informed or blind conditions (Ittyerah 1993, 2000, 2009), then the expected asymmetry between the hands (Annett 1985) is not evident and the blind may perform as well as the sighted with reliable self referent cues (Millar 1981, 1994) and gain in proficiency in the haptic processing of visuospatial information.

An absence of differences between the hands both with and without practice, indicates an equally good performance with both hands in the total absence of vision for prehensile movements that involve sorting and stacking of objects, the fine coordination of the thumb and forefinger as in finger dexterity tasks and the general ability of the fingers of both hands in the manipulation tasks. Therefore vision does not affect the general maturation of the child since the blind can gain in proficiency with practice of visuospatial tasks in the total absence of vision. This proficiency is not only confined to the preferred hand but is also to the same extent in the nonpreferred hand. Findings indicate that general laterality does not affect ability and suggest equipotentiality between the hands for both prehensile and nonprehensile actions. This is in conformity with evidence that indicates lesser discrepancies between the hands with age in children (Roy, Bryden & Cavill 2003) and little asymmetry between the hands for complex tasks (Hausmann, Kirk & Corballis 2004) because complex tasks are controlled by more widely distributed neuronal asymmetries in both hemispheres. Bush and Martin (2004) found no intermanual differences in healthy adults on the Rey complex figure test. Derakshan (2003) suggested that lateralities of movement in humans are controlled by the nature of callosal traffic in each hemisphere and not by the dominant hemisphere. These findings consequently strengthen the present theoretical notion of equal hand ability and suggest that hand ability may arise as a consequence of the callosal fibres that control functions in each hand and not the earlier notions of hemispheric dominance that were expected to control hand actions.

5.4 Implications of hand ability

The general conclusion from the central thesis of the manuscript is that vision does not determine hand preference and therefore hand actions are not solely dependent upon eye- hand coordination (e.g. Honda 1984; von Hofsten 1982; Jeannerod 1994, 1997a, 1997b). Furthermore, there is little or no relation between the preferred hand and proficiency in actions of skill. Hand preference is not an indication of skill and there is no effect of the hand used in the task (Ittyerah 1993, 2000, 2009). One hand may not be inherently more skilled than another, but rather task demands dictate hand actions (Millar 1987). Thus future directions of research relating preference and skill need explore the role of both hands in actions of skill.

Previous attempts to assess hand ability with tests of manual dexterity have indicated that these tests are good predictors of tactile task requirements such as type writing and pre-engineering skills for sighted and visually handicapped people (Tobin & Hunter 1974). Tobin and Hunter (1974) found that the Purdue pegboard test that assessed finger dexterity (by requiring subjects to insert pins into holes) and assembly (test that required assembling collars and washers) was performed fairly successfully by the visually handicapped with differing degrees of blindness such as the totally blind, light perception, residual vision and read ink print groups with the left as well as the right hands. Though the scores of the visually handicapped were almost half that of the sighted, their scores on the Purdue pegboard correlated with teacher ratings of aptitude for typewriting and pre- engineering skills. These findings can be considered to be of utility for vocational guidance and placement particularly for blind individuals. Apart from the practical values of assessment, training and rehabilitation, such effects may also contribute to the understanding of inter-relationships among various nonverbal aptitudes that are associated with success in specific training and job situations.

Recent evidence indicates that motor performance may remain steady with increasing age in the nonpreferred hand controlled by the right hemisphere. Poston, Enoka and Enoka (2008) quantified the aging related differences in end-point accuracy during isometric contractions of the left and right hands. It was expected that declines in motor performance with aging may be greater for muscles controlled by the right hemisphere. Both young (21–30) and old (76–80) men performed a task that involved matching the peak of a force time trajectory to a target. The old were less accurate than the young and exhibited greater end point error with the left than the right hand initially. Although myographic amplitude was similar between the hands, old adults exhibited greater timing variability. Given sufficient practice, there was no difference in endpoint accuracy between the left and right hands of old adults. This finding is not consistent with the prediction of an asymmetric decline in motor performance by the right hemisphere aging model. Furthermore recent evidence indicates that the hands may not vary in accuracy for proprioceptive localization. Jones, Cressman and Henrique (2010) required subjects to judge the position of a target hand relative to a visual reference or their midline body axis. Subjects also reached to the unseen location of a hand. Both left and right hands served as the target hand and the reaching hand. Subjects were good at estimating the location of each hand in both instances. Subjects perceived their left hand to be more to the left and their right hand more to the right of the reference point. The precision of estimates did not differ between the hands or vary with target location.

There is also evidence (Degrave, Brenner & Smeets 2009) that the Bretano illusion, which is a variation of the Muller- Lyer illusion, influenced goal directed movements of the left and right hands to the same extent irrespective of the visibility of the hand, such that there are no fundamental differences between the hands in susceptibility to the Bretano illusion.

Subitizing items by touch or tactile enumeration is similar for each hand. Plaiser (2009a) demonstrated that subitizing items by touch with the hands was possible just as for vision and both unimanual and bimanual subitizing indicated that subitizing ends for items held in the hand that are more than three and counting begins for items more than four. However, items were subitized regardless of their shapes. Items are subitized regardless of their homogeneity (only cubes) or heterogeneity of shape (cubes and spheres) and this is different from item individuation that searches for features of items to tell a sphere from a cube. Each hand was able to subitize a total of three items showing that the ability for tactile enumeration does not differ between the hands. Ferrand, Riggs and Castronova (2010) investigated the role of vision in tactile enumeration within and outside the subitizing range. Congenitally blind and sighted blindfolded subjects were required to enumerate quickly and accurately the number of fingers stimulated. Both groups enumerated 1 to 3 fingers quickly, but were slower with 4 to 9 fingers. Within the subitizing range, the blind were like the sighted blindfolded and seeing subjects. Outside the subitizing range, the blind and seeing subjects were better than the sighted blindfolded, suggesting that lack of access to the predominant sensory modality does affect performance. Subitizing is a general perceptual mechanism and vision is not necessary for the development of the subitizing mechanism.

Instances of an absence of performance difference between the hands in sighted subjects, have been found for a key pressing task (Ja'gr 1991), spatial accuracy following practice for long and short movements (Sherwood 1994) and drawing circles using symmetrical movements (Semjen, Summers & Cattaert 1995). Roy, Bryden and Cavill (2003) reported that the early preferred hand advantages found in 5 year olds for a peg moving task may be overcome during development in adults as a consequence of greater changes in performance with the non preferred hand as a function of age. For example, Bush and Martin (2004) found no intermanual differences in healthy adults with the dominant and non dominant hand for the Rey Complex figure test. Hausman, Kirk and Corballis (2004) suggested that for simple tasks such as finger tapping, the preferred hand may be faster because simple motor tasks involve localized neural networks confined to one hemisphere, whereas for more complex tasks the asymmetry diminishes because complex hand tasks may be controlled by more widely distributed neuronal asymmetries in both hemispheres.

There are suggestions that developmental models in cerebral maturation should examine the rate of improvement of left-sided skill in relation to right-side skill. Although hand preferences emerge early in development (Bryden, Pryde & Roy 2000; Oeztuerk et al. 1999; Pryde, Bryden & Roy 2000) between the ages of 2 and 5 years, performance differences between the hands gradually diminish after the age of 5 years partly due to the rapid development and myelination of the corpus callosum that allows for more efficient communication from the dominant to nondominant motor cortex (Barnea-Goraly et al. 2005; Denckla 1973). The emergence of hand preference is expected to accompany asymmetries in motor speed and coordination. In a finger tapping task Denckla (1973) found that right-handed finger repetition was faster than left-handed finger repetition in typically developing 5- to 7-year-old children. While the percentage of children performing faster with the right hand increased with age, the magnitude of this right-hand superiority decreased with age. Though lateral hand preference was evident during the elementary school years the non dominant hand speed/efficiency also improved alongside the preferred hand.

The performance ability of each hand indicates some effects of neurodevelopment such as the mylenation and development of the corpus callosum that allow for information from one hemisphere to be transferred to the other hemisphere. There is some speculation that motor skill development, particularly the developmental pattern of asymmetries in left- versus right-sided performance, may be a marker for maturation of the corpus callosum or for the different rates of development of the cerebral hemispheres (Roeder et al. 2008). Performance differences between preferred- and nonpreferred-sided activities were observed to follow a developmental trajectory in which there are initially small performance differences during infancy, followed by differences in favour of a developing preferred side (preschool through adolescence; De Bellis et al. 2001), and furthermore a reduction in left- right performance differences or near equalization of right and left skills after 9 years for repetitive finger tapping, sets of finger sequences, and repetitive foot taps (Denckla 1985). Roeder et al. (2008) tested typically developing children without any disability between the ages of 7 and 14 years on psychometric and performance tasks and observed that left side- right side differences diminished with age on timed unimanual and unipedal tasks. The authors concluded that the reduction of left- versus right-side speed on timed motor activities from 7 to 14 years of age is considered to reflect the myelination and maturation of cortical brain systems supporting motor skill, as well as the rapid myelination of the corpus callosum.

These findings also suggest that it may not be appropriate to assume functional representation solely from hand advantages because hand advantages might depend on the type of task and relevant factors that the task demands (Millar 1984, 1987). Millar and Al-Attar (2003) have for example demonstrated that the left and right hands do not differ in a spatial processing task that assessed distance. Thus comparisons between the hands in the processing of sensory inputs (Millar & Al-Attar 2003a) must be considered separately from comparisons between the hands for spatial versus sequential processes (Langdon & Warrington 2000). Processing of sensory inputs from touch and movement is not expected to differ between the hands.

Equal hand ability in performance tasks can be attributed to the fact that the right hand performs better for movement and the left hand for processing space (Kimura 1993). Neurological evidence indicates that the right hemisphere is more involved in spatial processing and the left hemisphere is involved with movement or sequential information and that both deal more with inputs from the opposite side of the body than ipsilateral inputs. Movements are mainly processed in the left cerebral hemisphere (De Renzi, Faglioni, Ledesani & Vecchi 1983; Kimura

1993) whereas the right hemisphere is dominant for processing spatial information from the opposite side of the body (Milner & Taylor 1972). Evidence indicates better recognition of wire shape with the left hand after cerebral comissurotomy (Kumar 1977). Thus the hands complement each other during haptic inspection and processing of stimuli.

Streri (2005) argued that hand ability is evident early in infancy so far as each hand may demonstrate particular competencies to detect features in complex objects. For example, Pineau and Streri (1990) found that 5-month-old infants were able to detect an unlikely element inside a complex object with their right hand. This showed that infants were able to process the contour of an object as well as the details inside the object. In order to demonstrate a manual specialization in global versus analytical information processing, 4-month-old babies participated in a tactile discrimination task between two objects: a square with concave or convex features and a disc with the same type of features. The "global" group had to differentiate the contour (the features are invariant) whereas the "detail" group had to differentiate the concave or convex aspect of the elements situated on the surface of the objects (the contour does not change). The right and left hands of each baby were stimulated. The babies seemed to prefer the right hand for discriminating features and the left hand for the contours (Streri 2002). Thus, when an object involving features on its surface was put in an infant's hands, a manual specialization (hemispheric specialization) was observed. Similar performances were found in visual spatial task processing. Four- to ten- month-old infants processed different information within geometrical patterns with each of two hemispheres (Deruelle & de Schönen 1991, 1995, 1998). The results of these studies have revealed that the left hemisphere has an advantage in processing the components of visual patterns whereas the right hemisphere has an advantage in processing the location of components of these patterns. Thus, hemispheric specialization is present very early in infancy both in the haptic and visual modes at the same age.

There is recent evidence that very early in infancy hand pointing responses may be correlated with speech or knowledge of spoken words. In a series of experiments (Jacquet, Esseily, Rider & Fagard 2012; Esseily, Jacquet & Fagard 2011) these authors demonstrated that the hand for grasping is independent of the hand for pointing. Infants as young as seven, fourteen and twenty months displayed more frequency of pointing with their right hands at an array of five targets that were displayed across the ipsilateral and contralateral hemispaces from left to right. When pointing at targets was compared to grasping targets, the frequency of grasping was less than pointing for both the left and right hands. Infants would more often grasp an object on the left side with their left hand than grasp an object on the right side with the left hand. Similarly grasping with the right hand was more frequent for targets on the right side than the left side. Whereas when pointing at targets, the right hand would point at targets both in the ipsilateral and contralateral spaces in all right handed children with a higher frequency as compared to grasping targets. Infants pointed to the leftmost target as often with their right hand as with their left hand and to the less leftward target more often with their right hand than the left hand. Furthermore, the frequency of right handers was higher for pointing than grasping and pointing with the right hand was associated with language development. Infants knowledge of spoken or produced words were higher for those that pointed with their right hands, whereas there was no association of word knowledge with the grasping response. These findings indicate a link between pointing and language development. They suggest that communicative gesture lateralization may be controlled by the left hemisphere for speech and is not a mere consequence of object- related handedness.

In a subsequent study Meunier, Vauclair and Fagard (2012), demonstrated the role of gestures in the origin of language by testing hand preferences to pointing or grasping at targets in several positions in space in human infants and adult baboons. The authors found a convergence in the hand preferences of the infants and the baboons to pointing and grasping. There were more right hand preferences for pointing at targets for both infants and baboons than for grasping. Therefore the communicative task generated more preferences for pointing particularly with the right hand suggesting that left lateralized language may have been initiated from gestural communication that was present across phylogeny and the early ancestors of man. These tests of pointing and grasping indicate that motor actions of grasping may be performed equally well with each hand, whereas hand gestures that indicate communication may be controlled by the left hemisphere. These findings reveal dissociation between motor acts performed by the hands and acts that involve a verbal component. All this evidence indicates equipotentiality between the hands for motor actions. Therefore theories of handedness that propose 'inferior-superior" (e.g. Annett 1985; Peters 1980) or 'just different' (Healey et al. 1986; Steinhuis & Bryden 1989, 1990) distinctions between the hands may not suffice since other evidence (Guiard 1987) argue for a complementary role between the hands as is common in daily acts of bimanual skills. In such instances usually one hand may be more dominant in a particular action, but there is a need for each hand to complement the other since the physical characteristics of movement and the sensory control mechanisms supporting each is different. Guiard argued that since the movements of the dominant hand tend to be lower in exertion, it can execute high frequency tasks (involving fine temporal and spatial resolution), while the other hand performs low frequency tasks (supporting an object). Rather than one hand being dominant, the complementary role differentiation model recognizes that both hands have different but equally important roles (Corbetta &

Thelen 1996). However, we (Ittyerah, Gaunet & Rossetti 2007) found that pointing at proprioceptive targets in the absence of vision revealed that differences between the hands were in their orientation during performance and not in ability. The right preferred hand adopted a context orientation during performance whereas the left non preferred hand adopted an egocentric orientation. Thus the difficulty during performance or increase in error usually observed for the left non preferred hand was a consequence of its egocentric orientation during performance and not a lack of ability.

Recent evidence reveals plasticity of human handedness in expert basket ball players (Stockel & Weigeit 2012). Basket ball players often have to adapt their skills to changes in play that require the flexible execution of a particular movement skill with either hand. Stockel and Weigeit compared professional and amateur players for the use of their non dominant hand during play to know the effects of proficiency change with practice. The authors observed that the use of the right dominant hand bias in basket ball players was reduced with increasing expertise. Professional players used their nondominant hand more often and with greater success than the less experienced players. Increasing amounts of bilateral practice can change task specific manual preference toward a greater use of both hands in competition and increase the proficiency of the nondominant hand.

The need for bilateral training was recognized as early as in the nineteenth century when Brown-Sequard (1874a, 1874b) suggested that if the inferior right hemisphere was educated as well as the superior left hemisphere, the consequences could produce a new superior intellectual being. During that time the left hemisphere was considered to be the larger of the two primarily due to a function of use. It was therefore proposed that if the left side of the body was used more often, the right hemisphere could develop like the left in size and function. Brown-Sequard recommended bilateral training as a means of stimulating both hemispheres by requiring every child as early as possible to exercise both sides of the body for writing, cutting or putting a spoon in the mouth so that the brain and spinal cord could develop while all their motor and sensitive elements were exercised. Thus, the potential for equal hand ability had long been recognized, though its scientific evidence is sparse mainly because of influential theories of hand dominance that have overlooked performance of the nondominant hand.

Although population studies of human handedness have revealed a right hand preference for most individuals, handedness at the individual level has indicated instances of preferred left hand use for certain actions, or performing certain actions more often with one hand (combing hair) than the other (hitting with a racquet/bat). Particularly skills of writing or eating have been encouraged to be performed with the right hand more often in eastern than western populations indicating that hand preferences may have emerged from cultural preferences. In India, the right hand is usually considered to be the good hand and often used for serving food and lighting the lamp during ceremonial occasions. But India also has a large population of artisans that employ both hands for particular occupations, indicating evidence for consistent motor ability in both hands. The best examples of equal hand ability can be observed among artisans or skilled handicraft workers. In India handicraft skill is usually a tradition that has been passed on to successive generations. India has a wide range of fine handicrafts in metal, stone, wood, handloom and motif designs ranging from Ladak in the north near the Himalayan mountain ranges to Kerala in the south at the tip of Cape Comorin. A fascinating feature of the Indian industrialization process is the co existence of traditional handicraft and modern mechanized production (Subramanian 2006).

Though art and craft is varied and predominantly traditional in different parts of India, I shall take examples of hand craft from Kerala, a southern state of India with which I am relatively more familiar. Pavakathakali or the art of performing with glove puppets worn on the hands and manipulated to perform artforms is one of the closest examples of the synchrony of the hand and brain. Kathakali, a traditional classical dance form in Kerala that usually depicts tales from Indian mythology is enacted with glove puppets. Traditional puppeteers have over generations sustained this deft skill. Unlike puppets that are tied to a string and manipulated as desired, the puppeteers of the Pavakathakali have no strings tied to their puppets. Instead the puppets are worn like gloves over the hands and the puppetry is conducted with three fingers in both hands. What is more than fascinating is that the thumb, the index and the middle fingers of both hands of the puppeteers are used to move the puppets, their heads, hands, the trunk or legs, while the arms of the puppeteers are held in a horizontal position with the puppets worn like gloves in their hands. The puppetry is performed on stage and is accompanied by song and rhythm. The emotions of the puppets are conveyed by the movement of the fingers of the puppeteers in synchrony with the song and rhythm. With the thumb, the index and the middle fingers, the art of storytelling in the Pavakathakali form is revived and made alive on stage. The manipulation of the fingers about the glove puppet, convey the mind of the protagonist as exposed by the puppeteer. There is every form of emotion and movement as can be seen in the usual stage performance with live actors. To be able to hold the arm horizontally for over an hour, while moving three fingers about the puppet is an art that skillfully demonstrates the equal ability of both hands.

The strings puppets, though not used in traditional art forms in Kerala are well used in Rajasthan, a state of Kings and Maharajas of the past in north- western India. The puppets are dressed in traditional bright Rajathani costumes and are manipulated by skilled puppeteers as they pull and loosen the strings tied to the head, neck, waist, arms and legs of the puppets. Sometimes the performance is fleeting and magical and conveys the deft skill of the fingers of both hands. The visual sensation could be one of puppets sliding over the floor of the stage, almost ethereal in all aspects. This is achieved with much practice on the part of the puppeteers, because an error in string pulling can bring the entire performance to a complete halt. Again the thumb and the forefingers are used to hold the strings as in a pincer grip, though the synchrony of the brain and the fingers reveal much versatility and connoisseurship.

Classical Indian dance like the Bharathanatyam not only generate movements of the feet of the artist, but is also accompanied by hand movements that maintain the rhythm of the performance. An important part of the hand movements is the formation of the fingers into symbols called 'mudras' that convey expression and thought. Some of the common mudras are the bird or the flower mudras that are formed with the thumb and the index fingers or a third finger where necessary. The dancer moves gracefully in rhythm with the music as s/he produces new mudras with her/his fingers to symbolize thought and unfold the message in the dance with movements of the feet and fingers. This is another example of equal hand ability with which the artist is able to evoke emotions from an audience.

The handloom weavers use their feet as well as their hands during the operation of the loom. Although it seems almost mechanical as the feet and the hands move in synchrony to set the loom in operation, both hands are used at the loom at the same time, one to adjust the shafts of the loom and the other to move the yarn shuttle to and fro. However, the work requires concentration so that fine cloth is produced. This is also evident in carpet weaving which seems more intricate with the colour designs woven along with the background of the carpet. In carpet weaving the fingers are often used to pick and move threads so that the desired designs appear on the carpet. These examples are a few, but sufficient to demonstrate the importance of hand actions in achieving particular goals. They raise questions about hand movements and hand skill involved in successful performance and the implications of hand ability.

The notion of equipotentiality between the hands is also evident in skills of juggling, playing musical instruments like the piano, typewriting and Braille reading in the blind. It shows that the intended movement to perform the skill can be executed with either hand. Even in actions of art and craft as in sculpturing or pottery making, both hands are constantly performing the required skilled movements. Each hand seems to complement the other showing a dependency for each other and its independent characteristics. Thus actions reflect thought, though little is scientifically known about the multiplicity of skilled hand actions that involve a variety of movement. If for example, the actions of living craftsmen were sustained and understood, they could possibly enlighten our understanding of the synchrony between hand and brain and the role of the hands in skilled performance. Artisans generate good data to support theoretical notions of equal hand ability that may be sustained with time.

Epilogue

The book is an exposition of the versatility of the human hands and its functional association with the sensory modality of touch. Though touch is the earliest modality to develop, it is least understood because of the variety of processes that are involved in knowing by touch. Even if each chapter deals with particulars and seems to take the reader away from the adjacent chapters, it is not easy to draw conclusions from any one of them without relating each to the other. Therefore the interconnectedness of the chapters must be appreciated.

The first chapter on Ways of Knowing, attempts to explain the interconnections among perception, action and language as important constituents of acquiring knowledge. These processes are not only confined to some modalities like vision or audition, by can also be understood in different ways by touch. Studies in tactile perception have shown that processes in touch are different from vision. Touch follows a serial process as the fingers of the hand move over the object to know its features, shape, size or texture. In vision, information is processed globally and quite simultaneously. While perceiving stimuli in the absence of vision, touch relies on an egocentric frame of reference. Usually the body mid-line is used as a reliable frame of reference to perceive object properties such as its shape or size. In sighted conditions, the observer relies on an allocentric frame of reference such as a wall or door, to know and compare object properties or relative distance between two objects. Perceptual processes in touch differ from those of vision or other sensory modalities.

Human beings are not passive recepients of stimulation, but are active and adaptive in their environments. As Jean Piaget had long observed, the child acts on the environment by assimilating and accommodating information during development. Studies with blind infants and children have shown that processes of assimilation and accommodation are not confined to vision, but are also present in the total absence of vision. Blind children have been found to have normal concepts of object permanence, perform better than sighted blindfolded children on Piagetians tests of seriation and are almost on a par with sighted children on tests of numerosity, space, and transitive thinking. Appropriate actions enable thought and development in the absence of vision.

Language is an important medium of communication and knowledge. Language can be verbal and assisted by speech or nonverbal as in instances of gestures and signs. Although the majority of people benefit from learning to speak and produce language verbally, and are able to read written text, language can also be effectively communicated and learnt from Braille texts or sign systems. Conceptual knowledge is attained from nonverbal linguistic systems and they must be recognized as appropriate ways of knowing.

As children mature they develop, but their cognitive development need not be dependent on vision, because information can be obtained from different sensory modalities. A popular view of cognitive development over the past few decades has been the theory of mind approach. Theorists favouring this approach have attempted to demonstrate that children understand the mental states of others by developing theories of the mind which reflect an innate capacity. Others have argued against the theory of mind approach to conclude that learning is an outcome of publicly shared knowledge such as communication between the mother and infant or listening to elders or peers. We (Ittyerah & Goyal 1997) have shown that blind children have an understanding of mental states in the total absence of vision, so long as they understand the concept. Thus the purpose of the chapter on Ways of Knowing is to generate awareness on the many ways of learning and knowing that are different from traditional practices of classroom instruction typically observed in sighted children.

Since the emphasis in the book is on ways of knowing by touch, the importance of the hand in attaining information must be clarified. Though touch encompasses the whole body, the focus of the manuscript is on the role of the hands in attaining information. Hand preference or handedness is a well known asymmetry and consequently an indication of lateralization. Since touching involves the hand, the hand and touch are intimately connected during acts of performance. The hands being controlled by the contralateral cerebral hemispheres, its evolution must be discussed to know the effects of lateralization across phylogeny.

Knowing about the evolution of hand use is important because it is a major contribution to the repertoire of human hand actions. Investigations of limb or foot use in mammals and birds have clarified ways of knowing lateral preference in nonhumans and enriched the understanding of lateralization in humans from comparative investigations. Findings of animal species tell of the evolution of vertebrate lateralization and how this is comparable to lateralization in humans. Of particular interest are the homologies in cerebral hemispheres that may exist between different animal vertebrate species and humans, which could reveal the role of the cerebral hemispheres in the control of actions on either side of the body. Studies in the evolution of lateralization in fishes, birds and mammals show that side preferences existed across phylogeny. Most of the studies show eye preferences or turning biases, whereas studies using paws or limbs are rare because of the difficulty involved in designing experiments and instructing animals in haptic tasks. No animal species is comparable to the human with regard to haptics. Studies in haptics come mostly from human subjects and are important in knowing the performance of each hand that is controlled by the contralateral hemisphere. The hand is a sensitive part of the body, particularly at the finger tips where most fine tactile discriminations are performed. Attention must also be given to fundamental grips of the hand so that these actions are incorporated in the study of haptic perception. Another interesting phenomenon is that in contrast to other senses, motor skills play a major role in haptics. Eye lenses are accommodated and the ears are tuned in the direction of the sound source, but in the case of the hand, motor skills are important for both executing an action and also using the hand as a sensory device, for example, to scan a surface.

The chapter on the hand links hand and touch with ways of knowing. Further, the role of vision in hand performance is also being clarified. Is reaching or pointing guided by vision? Theories of eye- hand coordination have long emphasized that hand preferences are an outcome of visuo-motor control. Experimental treatments that assess the role of vision have shown that animals displayed modality specific deficits after periods of dark rearing, indicating critical periods in the development of particular visual behaviours. Among humans although restricted rearing conditions are considered unethical, certain clinical conditions that reproduce the circumstances needed to study the effects of experience on perception have been valuable. Along with vision, movement of the entire body is considered to be important for perceptual development. Stimulus input as a result of an individual's own movement is known as reafference, whereas stimulus input acting on a passive observer is known as exafference. Subjects were also observed to adapt themselves in perceptually rearranged environments with feedback from active movement.

The predominance of studies in reaching and grasping attribute successful performance to the ability of the visual modality for aligning the hand and arm to the size and orientation of the object. The role of vision is undoubtedly facilitating, but speculations as to whether it is a necessary modality in the attainment of these specific behaviours can only be examined by comparing congenitally blind children with sighted cohorts. Studies with infants have shown that early in infancy proprioception takes precedence over vision in reaching and grasping objects, whereas eye hand coordination develops only after the age of about five months. Therefore performance is not dependent on vision early in infancy, until vision dominates later. Studies of hand preference in blind and sighted blindfolded children revealed that hand preferences are not determined by vision. Hand preference refers to the choice of a preferred hand and this is different from ability. The potential ability to perform a task may be present in both hands and evidence indicates that the hand used to perform tactile tasks is independent of one's hand preference and vision does neither determine hand preference nor hand ability in blind or sighted conditions. What we know of the hand and the role of vision must also be related to processes of the hand and brain that control hand actions.

Processes of the hand and brain involve acts of perception and action. Actions of reaching to targets in space performed by sighted people can also be performed by congenitally blind subjects in the total absence of vision. It is now established that processes for knowing what an object is, is different from knowing where the object is. This dissociation has been repeatedly demonstrated in brain damaged and normal subjects. In these instances, the hand has been observed to reach the target without conscious processing of the shape or the direction of the target. This dissociation has also been demonstrated in pointing tasks with congenitally blind subjects indicating that processes in pointing do not depend on vision. The ability to point separately with each hand showed, that the hands do not differ in ability and the difference between the preferred and the nonpreferred hand is related to the orientation of the hand during performance and not in ability. Therefore hand and brain relationships reveal haptic processes involved with each hand.

Hand and brain relationships are also evident when the hand is used as a frame of reference in the haptic comparison of stimuli. This indicates the influence of the position of the hand during haptic tasks so that it is treated as a frame of reference as in the instance of the body midline. Furthermore in instances when a tool is held in the hand to retrieve an object, sensory neurons incorporate the tool as an extension of the hand. The field sizes of the neurons seem to elongate along the tool, the closer the hand is to the tool. Such brain behavior relationships cannot be studied in isolation since they involve the integration of the hand and the tactile modality. Another aspect of hand and brain behavior is lateralization. Although lateralization is a well studied behavior at the individual and the population levels, an absence of lateralization has indicated a prognosis for developmental delay. Crow (1998) emphasized children who did not show a hand preference in the behavior checklist tasks had a prognosis for schizophrenia in adulthood. An incongruence of hand and brain relationship may be a concern for mental health.

Tactile cognition is the recognition of what is perceived when an object is felt or held in the hand. The control of processes in the contralateral hemisphere is evident when the right hand performs better in acts of movement over the stimulus and the left hand performs better in the detection of spatial characteristic of the stimulus. Both processes contribute to tactile recognition and implicitly indicate the complementary roles of the hands in tactile cognition. Studies of crossmodal matching that involve holding or perceiving the stimulus with the hand and recognizing it visually are tests of how the characteristics of the stimulus in one modality is recognized in another modality. It also indicates that processes of intersensory integration are important to know about object characteristics and there is unity of the senses. Vision by itself may not provide all the stimulus information; therefore a multisensory approach in learning must be appreciated. Shape perception depends on spatial organization of the stimulus wherein the felt features have to be located by reference to each other or to some internal or external frame. The frontal cortex, more specifically the pre frontal areas are interconnected with the parietal lobes as well as with the subcortical areas and these are also involved in the representation of space.

The visual and the tactile sensory modalities differ in their sensitivity to shape perception. The visual modality is at least two d primes (standard deviations) more sensitive than the tactile modality while perceiving 3D objects such as curvatures. During touch perception the hands move over the object to perceive stimulus characteristics and rely on movement and spatial processes. Systematic exploration of stimulus characteristics with the hand or fingers requires an anchor or reference point that can be recognized as the end and starting point of the exploratory movement. To know about spatial processes independent of hand effects, Millar and Al-Attar (2003b) tested two hypotheses. The first hypothesis that the left hand is better for spatial tasks, predicts a left hand advantage for performance in all conditions. The alternate reference hypothesis predicts significantly greater accuracy in haptic recall with explicit additional reference information than in conditions that do not provide additional reference information. Findings showed no differences in task performance when the left hand was placed on an external reference around the task space while the right hand pursued the target or when the right hand was placed on an external reference while the left hand pursued the target, showing haptic cognition does not differ between the hands. Furthermore there is evidence that subitizing items indicating the perception of tactile numerosity does not differ between the hands. Each hand was able to subitize small 3D shapes up to three items regardless of the shapes of the stimuli, whether the item was a cube or a sphere. It also indicates that vision is not necessary for tactile cognition and performance does not suffer from accuracy in the total absence of vision. Vision can improve haptic performance only if it provides task relevant cues (Millar & Al-Attar 2005).

As the book reads from chapter to chapter, in an attempt to relate aspects of hand and touch, the reader will also be able to appreciate that the preferred hand is not always the skilled hand and handedness or lateralization is not the same as hand ability. If people without hands are able to perform art with their feet or people without arms or legs are able to use their mouth to hold implements to draw or sketch, it indicates that movement intent can be executed from an alternate part of the body and that the motor cortex can respond to the congenital absence of limbs. The notion that the preferred hand is the skilled hand is no longer acceptable; rather performance could depend on the way attention is distributed to each hand (Peters 1991b). Studies with amelics experiencing phantom limb sensations have shown the visual analyses of other's actions can influence their body schema. There is also evidence of motor imagery aiding the rotation of the missing hand among people with hand amelia. These examples of cerebral plasticity show that movement intent can be expressed flexibly even in the absence of limbs.

How do motor skills develop during childhood? There seems to be a relationship between neuromotor maturity and handedness (Bishop 1990a). Consistent hand preferences were related to an advanced developmental status. This raises the question as to whether children with unestablished handedness are less consistent in crossing the body midline or commit more mirror movements and are less able to perform bimanual asymmetric actions. Some evidence indicates that children that are not lateralized have a prognosis for mental disorders during adulthood. The gist is that random handedness or an absence of lateralization is an indication of low performance on intelligence or ability tests.

There are suggestions that developmental models in cerebral maturation should examine the rate of improvement of left-sided skill *in relation to* right-side skill. Although hand preferences emerge early in development between the ages of 2 and 5 years, performance differences between the hands gradually diminish after the age of 5 years partly due to the rapid development and myelination of the corpus callosum that allows for more efficient communication from the dominant to nondominant motor cortex. The emergence of hand preference is expected to accompany asymmetries in motor speed and coordination. In a finger tapping task, right-handed finger repetition was faster than left-handed finger repetition in typically developing 5- to 7-year-old children (Denckla 1973). While the percentage of children performing faster with the right hand increased with age, the magnitude of this right-hand superiority decreased with age. Though lateral hand preference was evident during the elementary school years the non dominant hand speed/efficiency also improved alongside the preferred hand.

Examples of equal hand ability have also been observed in congenitally blind conditions indicating that vision does not determine hand preference or hand ability and theories of eye-hand coordination are insufficient explanations of hand preference. Hand ability is evident early in infancy so far as each hand may demonstrate particular competencies to detect features in complex objects. Infants displayed more right hand preferences for pointing at targets that had a verbal component such as a word, than for grasping at objects. Therefore the communicative task generated more preferences for pointing particularly with the right hand suggesting that left lateralized language may have been initiated from gestural communication that was present across phylogeny and the early ancestors of man. These tests of pointing and grasping indicate that motor actions of grasping may be performed equally well with each hand, whereas hand gestures that indicate communication may be controlled by the left hemisphere. These findings reveal dissociation between motor acts performed by the hands and acts that involve a verbal component. The evidence indicates equipotentiality between the hands for motor actions as is common among artisans who use both hands during skilled performance such as pottery making or sculpture. Of recent there is evidence (Uomini 2009) for a complementary role differentiation between the hands as is common in daily acts of bimanual skills. In such instances usually one hand may be more dominant in a particular action, but there is a need for each hand to complement the other since the physical characteristics of movement and the sensory control mechanisms supporting each is different (Guiard 1997). According to Uomini, hand actions vary in their gradient of task complexity, beginning with simple actions such as placing flints on a puzzle being coordinated by both hands, followed by relatively more lateralized actions such as picking objects with the thumb and forefinger, and hand laterality is at the extreme of the gradient for complementary bimanual tasks such as nut cracking or retrieving food from a hose/tube like container among chimpanzees, as the non preferred hand is used to hold the tube while the preferred hand draws food from the container. These examples display the involvement of both hands in actions and do not confine performance only to the preferred hand. Studies on hand preference, hand ability and the relationship between the hand and brain continue to show that skill is not confined to the preferred hand and the hands do not differ in tactile cognition. Just as ways of knowing are not confined to the sensory modality of vision, the notion of equipotentiality of the hands must be appreciated.

References

Abravanel, E. (1971). Active detection of solid shape information by touch and vision. *Perception and Psychophysics, 10*, 358–360.

Adelson, E., & Fraiberg, S. (1974). Gross motor development in infants blind from birth. *Child Development, 45*, 114–126.

Adelstein, A., & Crowne, D. P. (1991). Visuospatial asymmetries and interocular transfer in the split-brain rat. *Behavioural Neuroscience, 105*, 459–469.

Akbarin, S., Berndl, K., Gruesser, O. J., Guidin, W. O., Pause, M., et al. (1988). Responses of single neurones in the parietalinsular vestibular cortex of primates. *Annals of the New York Academy of Sciences, 545*, 187–202.

Albe-Fessard, D. (1967). Organization of somatic central projections. In W. D. Neff (Ed.), *Contributions to sensory physiology, 2* (pp. 101–167). New York: Academic Press.

Alegria, J., & Noirot, E. (1978). Neonate orientation behaviour toward human voice. *International Journal of behavioural Development, 1*, 291–312.

Alibali, M. W., Kita, S., & Young, A. J. (2000). Gesture and the process of speech production: We think, therefore we gesture. *Language & Cognitive Processes, 15*, 593–613.

Amedi, A., Jacobson, B., Malach, R., & Zohary, E. (2002). Convergence of visual and tactile shape processing in the human lateral occipital complex. *Cerebral Cortex, 12*, 1202–1212.

Amedi, A., Malach, R., Hendler, T., Peled, S., & Zohary, E. (2001). Visuo-haptic object activation in the ventral visual pathway. *Nature Neuroscience, 4*, 324–330.

Ament, W. (1923). *The mind of the child: A comparative history of life*. London: Simpkin Marshall.

Anderson, E. M., & Spain, B. (1977). *The child with spina bifidia*. Metheun: London.

Andrew, R. J. (2002). The earliest origins and subsequent evolution of lateralization. In L. J. Rogers, & R. J. Andrew (Eds.), *Comparative vertebrate lateralization* (pp. 70–93). New York: Cambridge University Press.

Andrew, R. J. (1991). The nature of behavioural lateralization in the chick. In R. J. Andrew (Ed.), *Neural and behavioural plasticity. The use of the chick as a model* (pp. 536–554). Oxford: Oxford University Press.

Andrew, R. J. (1983). Lateralization of emotional and cognitive function in higher vertebrates with special reference to the domestic chick. In J. P. Ewert, R. R. Capranica, & D. J. Ingle (Eds.), *Advances in vertebrate neuroethology* (pp. 477–505). New York: Plenum Press.

Andrew, R. J., & Rogers, L. J. (2002). The nature of lateralization in tetrapods. In L. J. Rogers, & R. J. Andrew (Eds.), *Comparative vertebrate lateralization* (pp. 94–125). New York: Cambridge University Press.

Andrew, R. J., & Watkins, J. A. S. (2002). Evidence for cerenral lateralization for senses other than vision. In L. J. Rogers, & R. J. Andrew (Eds.), *Comparative vertebrate lateralization* (pp. 365–382). New York: Cambridge University Press.

Andrew, R. J., Tomassi, L., & Ford, N. (2000). Motor control by vision and the evolution of cerebral lateralization. *Brain and Language, 73*, 220–235.

Andrew, R. J., Mench, J., & Rainey, C. (1982). Right-left asymmetry of response to visual stimuli in the domestic chick. In D. J. Ingle, M. A. Goodale, & R. J. Mansfield (Eds.), *Analysis of visual behaviour* (pp. 197–209). Cambridge, MA: MIT Press.

Annett, M. (2002). *Handedness and brain asymmetry. The right shift theory*. London: Taylor & Francis, Psychology Press.

Annett, M. (1985). *Left, right, hand and brain: The right shift theory*. London: LEA.

Annett, M. (1978). Genetic and non genetic influences on handedness. *Behaviour Genetics, 8*, 227–249.

Annett, M. (1976). A coordination of hand preference and skill replicated. *British Journal of psychology, 67*, 587–592.

Annett, M. (1973). Handedness in families. *Annals of Human Genetics, 37*, 93–105.

Annett, M. (1972). The distribution of manual asymmetry. *British Journal of Psychology, 63*, 343–358.

Annett, M. (1970a). A classification of hand preferences by association analysis. *British Journal of Psychology, 61*, 303–321.

Annett, M. (1970b). The growth of manual preference and speed. *British Journal of Psychology, 61*, 545–558.

Annett, M. (1967). The binomial distribution of right, mixed and left handedness. *Quarterly Journal of Experimental Psychology, 29*, 327–333.

Annett, M., & Manning, M. (1990). Reading and a balanced polymorphism for laterality and ability. *Journal of Child Psychology and Psychiatryand Allied disciplines, 31*, 511–529.

Annett, M., & Kilshaw, D. (1983). Right and left hand skill II: Estimating the parameters of the distribution in L–R differences in males and females. *British Journal of psychology, 74*, 269–283.

Annett, J., Annett, M., Hudson, P. T. W., & Turner, A. (1979). The control of movement in the preferred and non preferred hands. *Quarterly Journal of Experimental Psychology, 31*, 641–652.

Annett, M., Hudson, P. T. W., & Turner, A. (1974). The reliability of differences between the hands in motor skill. *Neuropsychologia, 12*, 527–531.

Annett, M., Golby, C. W., & Kay, H. (1958). The measurement of elements in an assembly Task-The information output of the human motor system. *Quarterly Journal of Experimental Psychology, 10*, 1–11.

Appelle, S., Gravetter, F. G., & Davidson, P. W. (1980). Proportion judgments in haptic and visual form perception. *Canadian Journal of Psychology, 34*, 161–174.

Arbib, M. A., Iberall, T., & Lyons, D. (1995). Coordinated control programs for control of the hand. In A. W. Goodwin, & I. Darian- Smith (Eds.), *Hand function and the neocortex* (pp. 135–170). Berlin: Springer Verlag.

Astington, J. W., Harris, P. L., & Olson, D. R. (1988). *Developing theories of mind*. New York: Cambridge University Press.

Aslin, R. N. (1993). Commentary. The strange attractiveness of dynamic systems to development. In L. Smith, & E. Thelen (Eds.), *A dynamic systems approach to development* (pp. 385–399). London: Bradford Books.

Ashton, G. C. (1982). Handedness: An alternative hypothesis. *Behaviour Genetics, 12*, 125–147.

Assaiante, C., & Amblard, B. (1992). Head-trunk coordination and locomotor equilibrium in 3 to 8 year old children. In A. Berthoz, P. P. Vidal, & W. Graf (Eds.), *The head- neck sensory motor system* (pp. 121–134). Oxford: Oxford University Press.

Atkinson, R. C., & Shiffrin, R. M. (1968). Human memory: A proposed system and its control processes. In K. W. Spence, & J. T. Spence (Eds.), *The psychology of learning and motivation*, Vol. 2 (pp. 89–195). Orlando, F L: Academic Press.

Attneave, F., & Benson, B. (1969). Spatial coding of tactual stimulation. *Journal of Experimental Psychology, 81*, 216–222.

Atweh, S. F., Banna, N. R., Jabbur, S. J., & To'met, G. F. (1974). Polysensory interactions in the cuneate nucleus. *Journal of Physiology, 238*, 343–355.

Baillargeon, R. (1994). Physical reasoning in infancy. In M. S. Gazzaniga (Ed.), *The cognitive neurosciences* (pp. 181–203). Cambridge, M A: MIT Press.

Bayley, N. (1969). *Bayley scales of infant development*. New York: Psychological Corporation.

Bard, C., Hay, L., & Fleury, M. (1990). Timing and accuracy of visually directed movements in children: Control of direction and amplitude components. *Journal of Experimental Child Psychology, 50*, 102–118.

Barnea-Goraly, N., Menon, V., Eckert, M., Tamm, L., Bammer, R., Karchemskiy, A., Dant, C.C., & Reiss, A. L. (2005). White matter development during childhood and adolescence: A cross- sectional diffusion tensor imaging study. *Cerebral Cortex, 15*, 1848–1854.

Barnsley, R. H., & Rabinovitch, M. S. (1970). Handedness: Proficiency versus stated preference. *Perceptual and Motor Skills, 30*, 343–362.

Bates, E., Marchman, V., Thal, D., Fenson, L., Dale, P., Reznick, S., Reilly, J., & Hartung, J. (1994). Development and stylistic variation in the composition of early vocabulary. *Journal of Child Language, 21*, 85–123.

Batheja, M., & McManus, I. C. (1985). Handedness in the mentally handicapped. *Developmental Medicine and Child Neurology, 27*, 63–68.

Battaglia- Mayer, A., Ferraina, S., Genovisio, A., Marconi, B., Squatriro, S., et al. Lacquaniti, F., & Caminiti, R. (2001). Eye hand coordination during reaching. An analysis of the relationships between visual manual signals in the parietal cortex and parieto- frontal association projection. *Cerebral Cortex, 11*, 528–544.

Bauer, R. H. (1993). Lateralization of neural control for vocalization by the frog (Rana pipiens). *Psychobiology, 21*, 243–248.

Bell, Charles. (1840). *The hand, its mechanism and vital endowments, as evincing design. The bridgewater treatises on the power, wisdom and goodness of god as manifested in creation.* (Treatise IV.) New York: Harper and Brothers.

Bellugi, U. (1988). The acquisition of a spatial language. In F. Kessel (Ed.), *The development of language and language researchers. Essays in honour of Roger Brown* (pp. 153–185). Hillsdale, New Jersey: LEA.

Bellugi, U. (1991). The link between hand and brain: Implications from a visual language. In D. S. Martin (Ed.), *Advances in cognition, education and deafness* (pp. 11–35). Washington DC: Gallaudet University Press.

Benton, A. L. (1985). Visuoperceptive, visuospatial and visuoconstructive disorders. In K. M. Heilman, & E. Valenstein (Eds.), *Clinical neuropsychology* (pp. 151–185). New York: Oxford University Press.

Benton, A. L., Varney, N. R., & Hamsher, K. de S. (1978). Lateral differences in tactile directional perception. *Neuropsychologia, 16*, 109–114.

Benton, A. L., Levin, H. S., & Varney, N. R. (1973). Tactile perception of direction in normal subjects. *Neurology, 23*, 1248–1250.

Bender, M. (1945). Extinction and precipitation of cutaneous sensations. *Archives of Neurology and Psychiatry, 54*, 1–9.

Berkley, G. (1709/1996). *An essay towards a new theory of vision*. London: Everyman Vision.

Berla, E. P. (1982). Haptic perception of tangible graphic displays. In W. Schiff, & E. Foulke (Eds.), *Tactual perception: A source book*. Cambridge: Cambridge University Press.

Berthenthal, B. I., & Bai, D. L. (1989). Infant's sensitivity to optic flow controlling posture. *Developmental Psychology, 25*, 936–945.

Berthoz, A. (1991). Reference frames for the perception and control of movement. In J. Paillard (Ed.), *Brain and space* (pp. 81–111). Oxford: Oxford University Press.

Berla, E. P., & Butterfield, L. J. Jr. (1977). Tactual distinctive feature analysis: Training blind students in shape recognition and in locating shapes on a map. *Journal of Special Education, 11*, 336–346.

Berrebi, A. S., Fitch, R. H., Ralph, D. L., Denenberg, J. O., Friedrich, V. L., & Denengerg, V. H. (1988). Corpus calosum: Region specific effects of sex, early experience and age. *Brain Research, 438*, 216–224.

Beukelaar, L. J., & Kroonenberg, P. M. (1983). Towards a conceptualization of hand preference. *British Journal of psychology, 74*, 33–45.

Bianki, V. L. (1988). *The right and left hemispheres of the animal brain: Cerebral lateralization of function. Monographs in neuroscience*, Vol. 3. New York: Gordon and Breach.

Bianki, V. L. (1982). Lateralization of concrete and abstract characteristics analysis in the animal brain. *International Journal of Neuroscience, 20*, 75–90.

Bigelow, A. (1986). The development of reaching in blind children. *British Journal of Developmental Psychology, 4*, 355–366.

Bisazza, A., Cantalupo, C., Capocchiano, M., & Vallortigara, V. (2000). Population lateralization and social behavior: A study with sixteen species of fish. *Laterality, 5*, 269–284.

Bisazza, A., Rogers, L. J., & Vallortigara, G. (1998). The origins of cerebral asymmetry: A review of evidence of behavioural and brain lateralization in fishes, amphibians and reptiles. *Neuroscience and Biobehavioural Reviews, 22*, 411–426.

Bisazza, A., Cantalupo, C., Robins, A., Rogers, L., & Vallortigara, G. (1997). Pawedeness and motor asymmetries in toads. *Laterlaity, 2*, 49–64.

Bisazza, A., & Vallortigara, G. (1996). Rotational bias in mosquitofish (Gambusia hoolbrooki): The role of lateralization and sun-compass navigation. *Laterality, 1*, 161–175.

Bisazza, A., Cantalupo, C., Capocchiano, M., & Vallortigara, G. (1996). Right pawedness in toads. *Nature, 379*, 408.

Bishop, D. V. M. (1990a). *Handedness and developmental disorder*. Oxford: Blackwell Scientific Publications.

Bishop, D. V. M. (1990b). Handedness, clumsiness and developmental language disorders. *Neuropsychologia, 28*, 681–690.

Bishop, D. V. M., & Edmundson, A. (1987). Language impaired 4-year olds. Distinguishing Transient from Persistent Impairment. Journal of Speech and Hearing Disorders Vol. *52* 156–173.

Bisiach, E., Capitani, E., Luzzatti, C., & Perani, D. (1981). Brain and conscious representation of outside reality. *Neuropsychologia, 19*, 543–551.

Bogen, J. E., Fisher, E. D., & Vogel, P. J. (1965). Cerebral commisurotomy: A second case report. *Journal of American Medical Association, 194*, 1328–1329.

Bogen, J. E., & Vogel, P. J. (1962). Cerebral commisurotomy in man. *Bulletin of Los Angeles Neurological Society, 27*, 165–172.

Bonda, E., Petrides, M., Frey, S., & Evans, A. 1995. Neural correlates of mental transformations of the body-in-space. In: *Proceedings of the National Academy of Sciences of the USA, 92*, 11180–11184.

Bossom, J., & Ommaya, A. (1968). A visuomotor adaptation to prismatic transformation of the retinal image in monkeys with bilateral dorsal rhizotomy. *Brain, 91*, 161–172.

Bousfield, W. A., Cohen, B. G., Whitmarsh, G. A. (1958). Associative clustering in the recall of words of different taxonomic frequencies of occurrence. *Psychological Reports, 4*, 39–44.

Boyd, J. (1967). Comparison of motor behaviour in deaf and hearing boys. *American Annals of the Deaf, 112*, 598–605.

Bower, T. G. R. (1974). *Development in infancy*. San Francisco: W. H. Freeman.

Brackenbridge, C. J. (1981). Secular variation in handedness over 90 years. *Neuropsychologia, 19*, 459–462.

Bradshaw, M. F., Watt, S. J., Elliott, K. M., & Riddell, P. M. (2004). The effects of a premovement delay on the kinematics of prehension in middle childhood. *Human Movement Science, 23*, 771–784.

Bradshaw, M. F., & Watt, S. J. (2002). A dissociation of perception and action in normal human observers: The effect of temporal delay. *Neuropsychologia, 40*, 1766–1778.

Bradshaw, J. L., & Rogers, L. J. (1993). *The evolution of lateral asymmetries, language, tool use and intellect*. New York : Academic Press.

Bradshaw, J. L., Nettleton, N. C., & Spehr, K. (1982). Braille reading and left and right hemispace. *Neuropsychologia, 20*, 493–500.

Brain, W. R. (1941). Visual disorientation with special reference to lesions of the right cerebral hemisphere. *Brain, 64*, 244–272.

Bremner, J. G. (1978). Spatial errors made by infants. Inadequate spatial cues or evidence of egocentrism? *British Journal of Psychology, 69*, 77–84.

Brennan, W., Ames, E. W., & Moore, R. W. (1966). Age differences in infants' attention to patterns of difference complexities. *Science, 151*, 354–356.

Bresson, F., Maury, L., Pieraut, Le., Bonniec, G., & de Shonen, S. (1977). Organization and lateralization of reaching in infants. An instance of asymmetric functions in hands collaboration. *Neuropsychologia, 15*, 311–320.

Brinkman, J., & Kuypers, H. G. J. M. (1973). Cerebral control of contralateral and ipsilateral arm, hand and finger movements in the split brain rhesus monkey. *Brain, 96*, 653–674.

Broadbent, D. E. (1956). Successive responses to simultaneous stimuli. *Quarterly Journal of Experimental Psychology, 8*, 145–152.

Broca, P. (1861). Perte de la parole. Rammollisment cronique et partielle du lobe anterieur gauche de cerveau. *Bulletin de la Societe d'Anthropologie, 2*, 235–237.

Bronchti, G., Heil, P., Sadka, R., Hess, A., Scheich, H., & Wollberg, Z. (2002). Auditory activation of "visual" cortical areas in the blind mole rat (*Spalaxehrenbergi*). *European Journal of Neuroscience, 16*, 311–329.

Brooks, V. B. (1984). 'How are "move" and "hold" programs matched?' In J. R. Bloedel, J. Dichgans, & W. Precht (Eds.), *Cerebellar functions* (pp. 1–23). Berlin: Springer.

Brown, J. A. (1958). Some tests of the decay theory of immediate memory. *Quarterly Journal of Experimental Psychology, 10*, 12–21.

Brown-Sequard, C. E. (1874a). Dual character of the brain. *Smithsonian Miscellaneous Collections, 15*, 1–21.

Brown-Sequard, C. E. (1874b). The brain power of man: He has two brains or has he only one? *Cincinnati Lancet and Observer, 17*, 330–333.

Bruner, J. S. (1977). Early social interaction and language acquisition. In H. R. Schaffer (Ed.), *Studies in mother-infant interaction* (pp. 271–289.). London: Academic Press.

Bruner, J. S., & Koslowski, B. (1972). Visually preadapted constituents of manipulatory action. *Perception, 1*, 3–14.

Brunt, D., & Broadhead, G. D. (1982). Motor profiency traits of deaf children. *Research Quarterly for exercise and sport, 53*, 236–238.

Bryant, P., Jones, P., Claxton, V., & Perkins, J. (1972). Recognition of shapes across modalities by infants. *Nature, 240*, 303–304.

Bryden, P. J., Pryde, K. M., & Roy, E. A. (2000). A developmental analysis of the relationship between hand preference and performance: II. A performance-based method of measuring hand preference in children. *Brain and Cognition, 43*, 60–64.

Buckhout, R. (1976). Eyewitness testimony. In R. Held, & W. Richards (Eds.), *Recent progress in perception* (pp. 205–213). San Francisco: Freeman.

Buckingham, G., Main, J. C., & Carey, D. P. (2011). Asymmetries in motor attention during a cued bimanual reaching task: Left and right handers compared. *Cortex, 47*, 432–440.

Budwig, N. (1989). The linguistic marking of agentivity and control in child language. *Journal of child language, 16*, 263–284.

Bullowa, M. (1979). *Before speech: The beginning of interpersonal communication.* Cambridge: Cambridge University Press.

Buneo, C. A., & Anderson, R. A. (2006). The posterior parietal cortex: Sensorimotor interface for the planning and online control of visually guided movements. *Neuropsychologia, 44*, 2594–2606.

Burton, H., Sinclair, R. J., & McLaren, D. G. (2004). Cortical activity to vibrotactile stimulation: An fMRI study in blind and sighted individuals. *Human Brain Mapping, 23*, 210–228.

Burton, H., Snyder, A. Z., Conturo, T. E., Akbudak, E., Ollinger, J. M., & Raichle, M. E. (2002). Adaptive changes in early and late blind: A fMRI study of Braille reading. *Journal of Neurophysiology, 87*, 589–607.

Bush, S., & Martin, T. A. (2004). Intermanual differences in the Rey complex figure test. *Rehabilitation Psychology, 49*, 76–78.

Bushnell, E. W., & Boudreau, J. P. (1991). The development of haptic perception during infancy. In M. Heller, & W. Schiff (Eds.), *The Psychology of touch* (pp. 139–161). Hillsdale, NJ: Lawrence Erlbaum Associates.

Butterfield, S. A., & Ersing, W. F. (1988). Influence of age, sex, hearing loss and balance on development of catching by deaf children. *Perceptual and motor skills, 66*, 997–998.

Butterfield, S. A., Mars- Hans, V., & Chase, J. (1993). Fundamental motor skills performances of deaf and hearing children ages 3 to 8. *Clinical Kinesiology, Spring*, 2–6.

Butterfield, E., & Siperstein, G. N. (1972). Influence of contingent auditory stimulation upon non-nutritive suckle. In J. Bosima (Ed.), *Third Symposium on oral sensation and perception.: The mouth of the infant.* Springfield, IL: Charles C. Thomas.

Butterworth, G., & Hicks, L. (1977). Visual, proprioceptive and postural stability in infancy. A developmental study. *Perception, 6*, 255–262.

Buyakas, T. M., Vardanyan, G., & Guppenreller, Y. B. (1980). On the mechanisms of precise hand movements. *Psichologitsheskij Journal, 1*, 93–103.

Caliskan, E., & Dane, S. (2009). Left handedness in blind and sighted children. *Laterality, 14*, 205–213.

Campbell, R. (1987). Lip reading and immediate memory processes or on thinking impure thoughts. In B. Dodd, & R. Campbell (Eds.), *Hearing by eye: The psychology of lip reading* (pp. 243–255). London: Lawrence Erlbaum.

Caplan, P. J., & Kinsbourne, M. (1976). Baby drops the rattle. Asymmetry of duration of grasp by infants. *Child Development, 47*, 532–534.

Carey, D. (2000). Eye hand coordination. Eye to hand or hand to eye? *Current Biology, 10*, 416–419.

Carey, S., & Diamond, R. (1977). From piecemeal to configurational representation of faces. *Science, 195*, 312–314.

Carlier, M., Doyen A-L., & Lamard, C. (2006). Midline crossing: Developmental trend from 3 to 10 years of age in a preferential card-reaching task. *Brain and Cognition, 61*, 255–261.

Carmon, A., & Benton, A. L. (1969). Tactile perception of direction and number in patients with unilateral cerebral disease. *Neurology, 19*, 525–532.

Carter, A. (1978). From sensori-motor vocalizations to words: A case study of the evolution of attention directing- communication in the second year. In A. Lock (Ed.), *Action, gesture and symbol: The emergence of language* (pp. 309–349). London: Academic Press.

Casperd, J. M., & Dunbar, R. I. M. (1996). Asymmetries in the visual processing of emotional cues during agnostic interactions in the gelada baboons. *Behavioural Processes, 37*, 57–65.

Castner, B. M. (1932). The development of fine prehension in infancy. *Genetic Psychology Monographs, 12*, 105–193.

Casteillo, U., Bennett, K. M. B., & Mucignat, C. (1993). The reach to grasp movement in blind subjects. *Experimental Brain Resarch, 96*, 152–162.

Casteillo, U., Paulignan, Y., & Jeannerod, M. (1991). Temporal dissociation of motor responses and subjective awareness. *A study in normal subjects. Brain, 114*, 2639–2655.

Chabot, N., Charbonneau, V., Laramée, M. E., Tremblay, R., Boire, D., & Bronchti, G. (2008). Subcortical auditory input to the primary visual cortex in anophthalmicmice. *Neuroscience Letters, 433*, 129–134.

Chance, S. A., & Crow, T. J. (2007). Distinctively human: Cerebral lateralization and language in homo sapiens. *Journal of Anthropological Sciences, 85*, 83–100.

Chance, S. A., Casanova, M. F., Switala, A. E., & Crow, T. J. (2006). Minicolumnar structure in Heschl's gyrus and planum temporale: Asymmetries in relation to sex and callosal fibre number. *Neuroscience, 143*, 1041–1050.

Chu, C. C., Tranel, D., & Damasio, H. (1994). How reliable are occipital asymmetry measurements? *Neuropsychologia, 32*, 1503–1513.

Church, R. B., Ayman-Nolley, S., & Mahootian, S. (2004). The role of gesture in bilingual education: Does gesture enhance learning? *International Journal of Bilingual Education and Bilingualism, 7*, 303–319.

Clayton, N. (1993). Lateralization and unilateral transfer of memory in marsh tits. *Journal of Comparative Physiology A, 171*, 799–806.

Clifton, R. K., Muir, D. W., Ashmead, D. H., & Clarkson, M. G. (1993). Is visually guided reaching in early infancy a myth? *Child development, 64*, 1099–1110.

Cohen, Leslie B. (1998). An information-processing approach to infant perception and cognition. In Francesca Simion, & George Butterworth, (Ed.), *The development of sensory, motor and cognitive capacities in early infancy: From perception to cognition* (pp. 277–300). Hove, England: Psychology Press/Erlbaum (UK) Taylor & Francis.

Cohen, L. G., Weeks, R. A., Sadato, N., Celnik, P., Ishii, K., & Hallett, M. (1999). Period of susceptibility for cross-modal plasticity in the blind. *Annals of Neurology, 45*, 451–460.

Colby, C. L. (1998). Action-oriented spatial reference frames in cortex. *Neuron, 20*, 15–24.

Collins, R. L. (1981). On asymmetries exhibiting a near equiprobable distribution of directions. *Behavioural Brain Sciences, 4*, 23–24.

Collins, R. L. (1975). When left handed mice live in right handed worlds. *Science, 187*, 118–184.

Collington, O., & De Volder, A. G. (2009). Further evidence that participants react faster to auditory and tactile spatial targets. *Canadian Journal of Experimental Psychology, 63*, 287–293.

Connolly, K. J. (1973). Factors influencing the learning of manual skills by young children. In R. A. Hinde, & J. S. Hinde (Eds.), *Constraints on learning* (pp. 337–365). London: Academic Press.

Connolly, K. J. (1970). Skill development: Problems and plans. In K. J. Connolly (Ed.), *Mechanisms of motor skill development* (pp. 3–17). London: Academic Press.

Connolly, K. J., & Elliot, J. M. (1972). Evolution and ontogeny of hand function. In N. Blurton-Jones (Ed.), *Ethological studies of child behaviour* (pp. 329–383). Cambridge: Cambride University Press.

Connolly, K., & Jones, B. (1970). A developmental study of afferent-efferent integration. *British Journal of Psychology, 61*, 259–266.

Cook, S. W., & Goldin-Meadow, S. (2006). The role of gesture in learning: Do children use their hands to change their minds? *Journal of Cognition and Development, 7*, 211–232.

Corballis, M. C. (2003). From mouth to hand: Gesture, speech and the evolution of right handedness. *Behavioural and Brain Sciences, 26*, 199–260.

Corkin, S. (1978). The role of different cerebral structures in somesthetic perception. In E. C. Carterette, & M. P. Friedman (Eds.), *Handbook of perception*, Vol. 6 (pp. 105–155). New York: Academic Press.

Corvette, D., & Thelen, E. (1996). The developmental origins of bimanual coordination: A dynamic perspective. *Journal of Experimental Psychology, 22*, 502–522.

Cowell, P. E., & Denenberg, V. H. (2002). Development of laterality and the role of the corpus callosum in rodents and humans. In L. J. Rogers, & R. J. Andrew (Eds.), *Comparative vertebrate lateralization* (pp. 274–305). New York: Cambridge University Press.

Cowell, P. E., Waters, N. S., & Deneberg, V. H. (1997). The effects of early environment in the development of functional laterality in Morris maze performance. *Laterality, 2*, 221–231.

Cozzutti, C., & Vallortigara, G. (2001). Hemispheric memories for the content and position of food caches in the domestic chick. *Behavioural Neuroscience, 115*, 305–313.

Cozzutti, C., & Vallortigara, G. (1999). Lateralization of a declarative memory in the chick as revealed by an incentive-devaluation procedure. Annual Meeting of the European Brain and Behaviour Society, Rome, 29 September-2 October, 1999.

Cratty, B. J. (1982). *Perceptual and motor development in infants and children*, 3rd edition. Engelwood Cliffs, N. J: Prentice Hall.

Crow, T. J. (2000). Schizophrenia as the price that homo sapiens pays for language: A resolution of the central paradox in the origin of species. *Brain Research Reviews, 31*, 118–131.

Crow, T. J. (1998). Nuclear Schizophrenic symptoms as a window on the transition between thought and speech. *British Journal of Psychiatry, 173*, 303–309.

Crow, J. J., Crow, L. R., Done, D. J., & Leask, S. (1998). Relative hand skill predicts academic ability: Global deficits at the point of hemispheric indecision. *Neuropsychologia, 36*(12), 1275–1282.

Crow, T. J., Done, D. J., & Sacker, A. (1996). Cerebral lateralization is delayed in children who later develop schizophrenia. *Schizophrenia Research, 22*, 181–185.

Crowne, D. P., Novotony, M. F., Mier, S. E., & Vitols, R. W. (1992). Spatial deficits and their lateralization following unilateral parietal cortex lesions in the rat. *Behavioural Neuroscience, 106*, 808–819.

Damerose, E., & Vauclair, J. (2002). Posture and laterality in human and nonhuman primates: Asymmetries in maternal handling and the infant's early motor asymmetries. In L. J. Rogers, & R. J. Richards (Eds.), *Comparative vertebrate lateralization* (pp. 306–362). New York: Cambridge University Press.

Davenport, R. K., & Rogers, C. M. (1970). Intermodal equivalence of stimuli in apes. *Science, 168*, 270–280.

Davenport, R. K., & Rogers, C. M. (1971). Perception of photographs by apes. *Behaviour, 39*, 318–320.

Davidson, P. W. (1972). The role of exploratory activity in haptic perception. Some issues, Data and hypotheses. *Research Bulletin, American Foundation for the Blind, 24*, 21–24.

De Bellis, M. D., Keshavan, M. S., Beers, S. R., Hall, J., Frustaci, K., et al. (2001). Sex differences in brain maturation during childhood and adolescence. *Cerebral Cortex, 11*, 552–557.

DeCasper, A. J., & Fifer, W. P. (1980). Of human bonding. Newborns prefer their mothers' voices. *Science, 208*, 1174–1176.

Deckel, A. W. (1995). Laterality of aggressive responses in Anolis. *Journal of Experimental Zoology, 272*, 194–200.

De Grave, D. D., Brenner, E., & Smeets, J. B. (2009). The Bretanno illusion influences goal directed movements of the left and right hands to the same extent. *Experimental Brain Research, 193*, 421–427.

Delantionio, A., & Riggio, L. (1981). Hemisphere lateralization and tactile modality. *Ricerch si psicologia, 5*, 275–310.

De Luca, J., Burright, R. G., & Donovick, P. J. (1990). Manual asymmetries during verbal and spatial block design construction. *Cortex, 6*, 541–542.

Denckla, M. B. (1985). Revised neurological examination for subtle signs. *Psychopharmacology Bulletin, 21*, 773–779.

Denckla, M. B. (1973). Development of speed in repetitive and successive finger-movements in normal children. *Developmental Medicine and Child Neurology, 15*, 635–645.

Denenberg, V. H. (1981). Hemispheric laterality in animals and the effects of early experience. *Behavioural and Brain Sciences, 4*, 1–49.

Denenberg, V. H., Gall, J. S., Berrebi, A., & Yutzey, D. A. (1978). Callosal mediation of cortical inhibition in the lateralized rat brain. *Brain Research, 397*, 327–332.

Derakhshan, I. (2003). Handedness, neural versus behavioural. Is there a measurable colossal difference? *Journal of Motor Behaviour, 35*, 409–411.

De Renzi, E. (1982). *Disorders of space exploration and cognition*. Chichester, Sussex: John Wiley.

De Renzi, E. (1978). Hemisphere asymmetry as evidenced by spatial disorders. In M. Kinsbourne (Ed.), *Asymmetrical function of the brain* (pp. 49–85). Cambridge: Cambridge University Press.

De Renzi, E., Faglioni, P., Lodesani, M., & Vecchi, A. (1983). Performance of left brain damaged patients on imitation of single movements and motor sequences: Frontal and parietal patients compared. *Cortex, 19*, 333–343.

De Renzi, E., & Spinnler, H. (1966a). Facial recognition in brain damaged patients. *Neurology, 16*, 145–152.

De Renzi, E., & Spinnler, H. (1966b). Visual recognition in patients with unilateral cerebral disease. *Journal of Nervous and Mental Disease, 142*, 515–525.

Deruelle, C., & Fagot, J. (1997). Hemispheric lateralization and global precedence effects in the Processing of visual stimuli by humans and baboons (Papio papio). *Laterality, 2*, 233–246.

Deruelle, C., & de Schonen, S. (1998). Do the right and left hemispheres attend to the same visuo-spatial information within a face in infancy? *Developmental Neuropsychology, 14*, 535–554.

Deruelle, C., & de Schonen, S. (1995). Pattern processing in infancy: Shape and locations of components are not processed by the same hemisphere. *Infant Behaviour and Development, 18*, 123–132.

Deruelle, C., & de Schonen, S. (1991). Hemisphereic asymmetries in visual pattern processing in infancy. *Brain and Cognition, 16*, 151–179.

De Saussure, F. (1916). *Course in general linguistics, Payot.* (Trans. R. Harris, 1983) La Salle, Illinois: Open Court, Paris.

De Vries, H. L. (1943). The quantum character of light and its bearing upon threshold of vision, the differential sensitivity and visual acuity of the eye. *Physica, 10*, 553–564.

Dodd, B. (1979). Lip reading in infants: Attention to speech presented in and out of synchrony. *Cognitive Psychology, 11*, 478–484.

Dodd, B. (1977). The role of vision in the perception of speech. *Perception*, 6, 31–40.

Dodds, A. G. (1978). Hemispheric differences in tactuo-spatial processing. *Neuropsychologia, 16*, 247–254.

Dodd, B., & Campbell, R. (Ed.). (1987). *Hearing by the eye. The psychology of lip reading.* London: Erlbaum.

Domellof, E., Fagard, J., Jacquet, A–Y., & Ro¨nnqvist, L. (2010). Goal-directed arm movements in children with fetal alcohol syndrome: A kinematic approach. *EFNS European Journal of Neurology*, doi:10.1111/j.1468–1331.2010.03142.x.

Domellof, E., Ro¨nnqvist, L., Titran, M., Rana Esseily, R., et al. (2009). A typical functional lateralization in children with fetal alcohol syndrome. *Developmental Psychobiology, 51*, 696–705.

Driver, J., Mattingley, J. B., Rorden, C., & Davis, G. (1997). Extinction as a paradigm measure of attentional bias and restricted capacity following brain injury. In P. Their, & H–O. Karnath (Eds.), *Parietal lobe contributions to orientation in 3D space* (pp. 401–429). Heidelberg: Springer.

Duncan, J. (1984). Selective attention and the organization of visual information. *Journal of Experimental Psychology, 113*, 501–517.

Duhamel, J. R., Colby, C. L., & Goldberg, M. E. (1991). Congruent representations of visual somatosensory space in single neurones of the monkey ventral intra-parietal cortex (Area VIP). In J. Paillard (Ed.), *Brain and space* (pp. 223–236). Oxford: Oxford University Press.

Dumas, C. (1992). Object permanence in cats (Felix catus): An ecological approach to the study of invisible displacements. *Journal of Comparative Psychology, 106*, 404–410.

Easton, R. D., Greene, A. J., & Srinivas, K. (1997). Transfer between vision and haptics. Memory for 2 D patterns and 3 D objects. *Psychonomic Bulletin and Review, 4*, 403–410.

Edin, B. B., Westling, G., & Johansson, R. S. (1992). Independent control of human finger-tip forces at individual digits during precision lifting. *Journal of Physiology, 450*, 547–564.

Ehret, G. (1987). Left hemisphere advantage in the mouse brain for recognizing ultrasonic communication calls. *Nature, 325*, 249–251.

Elimas, P. D. (1975). Auditory and phonetic coding of the cues for speech: Discrimination of the (r–l) distinction by young infants. *Perception and Psychophysics, 18*, 341–347.

Elimas, P. D. (1974). Auditory and linguistic processing of cues for place of articulation by infants. *Perception and Psychophysics, 16*, 513–521.

Elimas, P. D., Siqueland, E. R., Jusczyk, P., & Vigorito, J. (1971). Speech perception in infants. *Science, 171*, 303–306.

Elliott, D., Carson, R. G., Goodman, D., & Chua, R. (1991). Discrete versus continuous visual control of manual aiming. *Human Movement Science, 10*, 393–418.

Elliot, D., Calvert, R., Jaeger, M., & Jones, R. (1990). A visual representation and the control of manual aiming movements. *Journal of Motor Behaviour, 22*, 327–346.

Elliot, D., & Madalena, J. (1987). The influence of premovement visual information on manual aiming. *Quarterly Journal of Experimental Psychology, 39A*, 541–549.

Elliot, J. M., & Connolly, K. J. (1984). Classification of manipulative hand movements. *Developmental Medicine and Child Neurology, 26*, 283–296.

Elman, J. L., Bates, E. A., Johnson, M. H., Karmiloff-Smith, A., Parisi, D., & Plunkett, K. (1996). *Rethinking innateness: A connectionist perspective on development.* Cambridge, MA: MIT Press.

Engel, R., & Young, N. B. (1969). Calibrated pure tone audiograms in normal neonates based on evoked electroencephalographic responses. *Neuropaediatrie, 1*, 149–160.

Enright, J. T. (1995). The non-visual impact of eye orientation on eye hand coordination. *Vision Research, 35*, 1611–1618.

Esseily, R., Jacquet, A–Y., & Fagard, J. (2011). Handedness for grasping objects and pointing and the development of language in 14-month-old infants, laterality: Asymmetries of body. *Brain and Cognition, 16*(5), 565–585.

Estes, D., Wellman, H. M., & Woolley, J. D. (1989). Children's understanding of mental phenomena. In H. W. Reese (Ed.), *Advances in child development and behaviour.* San Diego, C. A: Academic Press.

Etienne, S. A. (1974). Age variability shown by domestic chicks in selected spatial tasks. *Behaviour, 50*, 52–76.

Fabre-Thorpe, M. M., Fagot, J., Lorincz, E., Levesque, F., & Vauclair, J. (1993). Laterality in cats: Paw preference and performance in a visuomotor activity. *Cortex, 29*, 15–24.

Fagot, J., & Deruelle, C. (1997). Processing of global and local visual information and hemispheric specialization in humans (Homo sapiens) and baboons (Papio papio). *Journal of Experimental psychology: Human Perception and Performance, 23*, 429–442.

Fagot, J., & Bard, K. A. (1995). Asymmetric grasping response in neonate chimpanzees (Pan troglodytes). *Infant Behaviour and Development, 18*, 253–255.

Falk, D., Cheverud, J., Vannier, M. W., & Conroy, G. C. (1986). Advanced computer graphics technology reveals aortical asymmetry in endocasts of rhesus monkeys. *Folia Primatologica, 8*, 1–40.

Fantz, R. L. (1965). Visual perception from birth as shown by pattern selectivity. *Annals of the New York Academy of Sciences, 118*, 713–814.

Fantz, R. L. (1961). The origin of form perception. *Scientific American, 204*(5), 66–72.

Fantz, R. L. (1958). Pattern vision in young infants. *Psychological Research, 8*, 66–72.

Farah, M. J. (1990). *Visual agnosia: Disorders of object recognition and what they tell us about normal vision.* Cambridge, MA: MIT Press.

Farne' A., & La`davas E. (2000). Dynamic size-change of hand peripersonal space following tool use. *Neuroreport, 11*, 1645–1649.

Fennel, E., Satz, P., & Wise, R. (1967). Laterality differences in the perception of pressure. *Journal of Neurology, Neurosurgery and Psychiatry, 30*, 337–340.

Ferrand, L., Riggs KJ., & Castronovo, J. (2010). Subitizing in congenitally blind adults. *Psychonomic Bulletin and Review, 17*(6), 840–845.

Ferrel-Chapus, C., Hay, L., Olivier I., Bard C., & Fleury, M. (2002). Visuo-manual coordination in childhood: Adaptation to visual distortion. *Experimental Brain Research, 144*, 506–517.

von Fersen, L., & Gunturkun, O. (1990). Visual memory lateralization in pigeons. *Neuropsychologia, 28*, 1–7.

Fertsch, P. (1947). Hand dominance in reading braille. *American Journal of Psychology, 60*, 335–349.

Fiehler, K., & Rösler, F. (2010). Plasticity of multisensory dorsal stream functions: Evidence from congenitally blind and sighted adults. *Restorative Neurology and Neuroscience*, *28*(2), 193–205.

Fine, M. L., McElroy, D., Rafi, J., King, C. B., & Loesser, K. E., et al. (1996). Lateralization of pectoral stridulation sound production in the channel catfish. *Physiology and Behaviour*, *60*, 753–757.

Fischer, K. W., & Corrigan, R. (1981). A skill approach to language development. In R. E. Stark (Ed.), *Language behaviour in infancy and childhood* (pp. 345–269). New York: Elsevier, North Holland.

Fix, J. D. (2002). *Neuroanatomy*, 3rd edition. Baltimore, MD: Lippincott, Williams, & Williams.

Flanders, M., Helms-Tillery, S. I., & Soechting, J. F. (1992). Early stages in a sensorimotor transformation. *Behavioral and Brain Sciences*, *15*, 309–362.

Fleishman, E. A. (1958). Dimensional analysis of movement reactions. *Journal of Experimental Psychology*, *55*, 438–453.

Fleishman, E. A., & Ellison, G. D. (1962). A factor analysis of fine manipulative tests. *Journal of Applied Psychology*, *46*, 96–105.

Flevares, L. M., & Perry, M. (2001). How many do you see? The use of nonspoken representations in first-grade mathematics lessons. *Journal of Educational Psychology*, *93*, 330–345.

Flowers, K. (1975). Handedness and controlled movement. *British Journal of Psychology*, *66*, 39–52.

Foley, J. E. (1974). Factors governing interocular transfer of prism adaptation. *Psychological Review*, *81*, 183–186.

Foley, J. E. (1970). Prism adaptation with opposed base orientation. The weighting of direction information from the two eyes. *Perception & Psychophysics*, *8*, 23–25.

Fontenot, D. J., & Benton, A. L. (1971). Tactile perception of direction in relation to hemispheric locus of lesion. *Neuropsychologia*, *9*, 83–88.

Forkman, B. (1998). Hens use occlusion to judge depth in a two-dimensional picture. *Perception*, *27*, 861–867.

Forssberg, H., Eliasson, A. C., Kinoshita, H., Westling, G., & Johansson, R. S. (1995). Development of human precision grip. (Tactile adaptation of isometric finger forces to the frictional condition). *Experimental Brain Research*, *104*, 323–330.

Forssberg, H., Eliasson, A. C., Kinishita, H., Johansson, R. S., & Westling, G. (1991). Development of human precision grip. I: Basic coordination of force. *Experimental Brain Research*, *85*, 451–457.

Foulke, E. (1982). Reading Braille. In W. Schiff, & E. Foulke (Ed.), *Tactual perception: A source book* (pp. 168–208). Cambridge: Cambridge University Press.

Fraiberg, S. (1977). *Insights from the blind*. London, UK: Souvenir Press.

Fraiberg, S. (1968). Parallel and divergent patterns in blind and sighted children. *Psychoanalytic study of the child*, *23*, 264–300.

Franco, L., & Sperry, R. W. (1977). Hemisphere lateralization for cognitive processing of geometry. *Neuropsychologia*, *15*, 107–114.

Friedman, D. A. (1971). Congenital and perinatal sensory deprivation: Some studies in early development. *American Journal of Psychology*, *127*, 1539–1545.

Friedman, H., & Davis, M. (1938). Left handedness in parrots. *Auk*, *80*, 478–480.

Funk, M., & Brugger, P. (2008). Mental rotation of congenitally absent hands. *Journal of International Neuropsychological Society*, *14*, 81–89.

Funk, M. (2006). *Cortical representations of body parts in children and individuals born with missing hands*. Unpublished Doctoral Dissertation, University of Zurich, Switzerland.

Funk, M., Shiffrar, M., & Brugger, P. (2005). Hand movement by observations by individuals born without hands: Phantom limb experience constraints visual limb perception. *Experimental Brain Research, 164,* 341–346.

Funk, M., & Brugger, P. (2002). Visual recognition of hands by persons born with only one hand. *Cortex, 38,* 860–863.

Furth, H. G. (1997). *Instinct and society.* Plenum Press.

Furth, H. G. (1994). The child's entry into the symbolic order. A developmental societal interpretation of J. Lacan. *Social Development, 3,* 158–184.

Galaburda, A. M. (1991). Asymmetries of cerebral neuroanatomy. In Wiley Chichester (Ed.), *Biological asymmetry and handedness* (Ciba Found Symp 162) (pp. 219–233).

Gannon, P. J., Broadfield, D. C., Kheck, N. M., Hof, P. R., & Braun, A. R., et al. (1988a). Brain language area evolution I: Anatomic expression of Heschl's gyrus and plannum temporal asymmetry in great apes, lesser apes, and old world monkeys. *Society for Neuroscience Abstracts, 24,* 160.

Gannon, P. J., Holloway, R. L., Broadfield, D. C., & Braun, A. R. (1988b). Asymmetry of chimpanzee planum temporal: Human like pattern of Wernicke's brain language area homolog. *Science, 279,* 220–222.

Gaunet, F., Ittyerah, M., & Rossetti, Y. (2007). Pointing at targets by children with congenital and transient blindness. *Experimental Brain Research, 178*(2), 167–179.

Gaunet, F., & Rossetti, Y. (2006). Effects of visual deprivation on representation: Immediate and delayed pointing toward memorized proprioceptive targets. *Perception, 35,* 107–124.

Gaydos, H. F. (1956). Intersensory transfer in the discrimination of form. *American Journal of Psychology, 69,* 107–110.

Gazzaniga, M. S. (1967). The split brain in man. *Scientific American, 217,* 24–29.

Gazzaniga, M. S., & Ledoux, J. E. (1978). *The integrated mind.* New York: Plenum Press.

Gazzaniga, M. S., & Sperry, R. W. (1967). Language after section of the cerebral commissures. *Brain, 90,* 131–148.

Gottfried, A. W., & Bathurst K. (1983). Hand preference across time is related to intelligence in young girls, not boys. *Science, 221,* 1074–1076.

Geschwind, N., & Galaburda, A. S. (1987). *Cerebral lateralization.* Cambridge. MA: MIT Press.

Geshwind, N., & Levitski, W. (1968). Human brain: Left-right asymmetry in the temporal speech region. *Science, 161,* 186–187.

Gessel, A. (1928). *Infancy and human growth.* New York: Macmillan.

Gessel, A. (1925). *The mental growth of the preschool child.* New York: Macmillan.

Gessel, A., & Amatruda, C. (1947). *Developmental diagnosis. Normal and abnormal child development.* New York: Harper & Row.

Gessel, A., & Ames, B. (1947). The development of handedness. *Journal of Genetic Psychology, 70,* 155–176.

Giachritsis, C., & Wing, A. (2010). The effect of bimanual lifting on grip force and weight perception. Haptics: Generating and perceiving tangible sensations. International conference, EuroHaptics 2010, Proceedings, Part 1.

Gibson, E. J. (1969). *Principles of perceptual learning and development.* New York: Appleton-Century-Crofts.

Gibson, J. J. (1966). *The senses considered as perceptual systems.* Boston, MA: Houghton Mifflin Compagny.

Gibson J. J. (1979). *The ecological approach to visual perception.* Boston: Houghton Mifflin.

Gibson, E. J., & Walk, R. D. (1960). The visual cliff. *Scientific American, 202*(4), 64–71.

Gibson, K. R., Rumbaugh, D., & Beran, M. (2001). Bigger is better: Primate brain size in relation to cognition. In D. Falk, & K. R. Gibson (Eds.), *Evolutionary anatomy of the primate cerebral cortex* (pp. 79–97). UK: Cambridge University Press.

Gilson, E. Q., & Baddeley, A. D. (1969). Tactile short term memory. *Quarterly Journal of Experimental Psychology, 21*, 180–184.

Glick, S. D., & Shapiro, R. M. (1985). Functional and neurochemical mechanisms of cerebral lateralization in rats. In S. D. Glick (Ed.), *Cerebral lateralization in non human species* (pp. 158–184). New York: Academic Press.

Goldin-Meadow, S., & Singer, M. A. (2003). From children's hands to adults' ears: Gesture's role in teaching and learning. *Developmental Psychology, 39*, 509–520.

Goldin-Meadow, S., Nusbaum, H., Kelly, S. D., & Wagner, S. (2001). Explaining math: Gesturing lightens the load. *Psychological Science, 12*, 516–522.

Goldin-Meadow, S., Kim, S., & Singer, M. (1999). What the teacher's hands tell the student's mind about math. *Journal of Educational Psychology, 91*, 720–730.

Goodale, M. A., Jacobson, L. S., Milner, A. D., Perret, D. I., & Benson, P. J., et al. (1994). The nature and limits of orientation and pattern processing supporting visuomotor and control in a visual form agnosic. *Journal of Cognitive Neuroscience, 6*, 46–56.

Goodale, M. A., & Milner, A. D. (1992). Separate visual pathways for perception and action. *Trends in Neuroscience, 15*(1), 20–25.

Goodale, M. A., Milner, A. D., Jacobson, L. S., & Carey, D. P. (1991). A neurological dissociation between perceiving objects and grasping them. *Nature, 349*, 154–156.

Goncalves, E., & Hoshino, K. (1990). Behavioural lateralization in the fresh water fish Oreochromis niloticus. Abstract translated from the *Annals of the 5th Annaula Meeting of the Federation of Brazilian Socieites for Experimental Biology (FESBE)*, Caxambu, M. G. p. 25.

Gordon, A. M., Westling, G., Cole, K. J., & Johansson, R. S. (1993). Memory representation underlying motor commands used during manipulation of common and novel objects. *Journal of Neurophysiology, 69*, 1789–1796.

Gordon, A. M., Forssberg, H., Johannson, R. S., Eliasson, A. C., & Westling, G. (1992). Development of human precision grip.3: Intergration of visual size areas during the programming of isometric forces. *Experimental Brain Research, 90*, 399–403.

Gordon, A. M., Forssberg, H., Johansson, R. S., & Westling, G. (1991). Visual size cues in the programming of manipulative forces during precision grip. *Experimental Brain Research, 83*, 477–482.

Gordon, H., W., & Kravetz, S. (1991). The influence of gender, handedness and performance level on specialized cognitive functioning. *Brain and Cognition, 15*, 37–61.

Gordon, I. A., & Morrison, V. (1982). The haptic perception of curvature. *Perception & Psychophysics, 31*, 446–450.

Gotestam, K. O. (1990). Left handedness among students of architecture and music. *Perceptual and Motor Skills, 70*, 1323–1327.

Gravede de D. D., Brenner E., & Smeets J. B. (2009). The Bretano illusion influences goal directed movements of the left and eight hand to the same extent. Experimental Brain Research, *193*, 421–427.

Graziano, M. S., & Gross, C. G. (1998). Spatial maps for the control of movement. *Current Opinion in Neurobiology, 8*, 195–201.

Graziano, M. S., & Gross, C. G. (1995). The representation of extrapersonal space: A possible role for bimodal, visual-tactile neurons. In M. S. Gazzaniga (Ed.), *The cognitive neurosciences*, (pp. 1021–1034). Cambridge, MA: MIT Press.

Green, D. M., & Swets, J. A. (1966). *Signal detection theory and psychophysics*. New York: Wiley.

Gregory, R. L., & Wallace, J. C. (1963). *Recovery from blindness: A case study, Experimental Psychology Society Monographs No. 2*. Cambridge: Cambridge University Press.

Greenfield, P. (1991). Language, tools and brain: The ontogeny and phylogeny of hierarchical organized sequential behaviour. *Behavioural and Brain Sciences, 14*, 531–595.

Grill-Spector, K., Kourtzi, Z., & Kanwisher, N. (2001). The lateral occipital complex and its role in object recognition. *Vision Research, 41*, 1409–1422.

Guan, J., & Wade, M. (2000). The effectof aging on adaptive eye hand coordination. *The Journals of Gerontology Series B: Psychological Science, 55*, 151–162.

Guiard, Y. (1987). Asymmetric division of labour in human skilled bimanual action: The kinematic chain as model. *Journal of Motor Behaviour, 19*(4), 486–517.

Gur, R., & Gur, R. (1977). Correlates of conjugate lateral eye movements in man. In S. Harnard, R. W. Doty, L. Goldstein, J. Jaynes, & G. Krauthamer (Eds.), *Lateralization in the nervous system* (pp. 261–281). New York: Academic Press.

Gurd, J. M., Schulz, S., Cherkas, L., & Ebers, G. C. (2006). Hand preference and performance in twenty pairs of monozygotic twins with discordant handedness. *Cortex, 42*, 934–945.

Halverson, H. M. (1937). Studies of grasping responses of early infancy: I, II, III. *Journal of Genetic Psychology, 51*, 371–449.

Halverson, H. M. (1932a). An experimental study of prehension in infants by means of systematic cinema records. *Genetic Psychology Monographs, 10*, 110–286.

Halverson, H. M. (1932b). A further study of grasping. *Journal of General Psychology, 7*, 34–63.

Halverson, H. M. (1931). An experimental study of prehension in infants by means of systematic cinema records. *Genetic psychology Monographs, 10*, 107–286.

Hamilton, C. R. (1967). Effects of brain bisection in eye hand coordination of monkeys wearing prism. *Journal of Comparative and Physiological Psychology, 64*, 434–443.

Hamilton, C. R., & Vermeire, B. A. (1998). Complementary hemispheric specialization in monkeys. *Science, 242*, 1691–1694.

Hamilton, C. R., & Vermeire, B. A. (1983). Discrimination of monkey faces by split- brain monkeys. *Behavioural Brain Reseach, 9*, 263–275.

Harley, R. K. Jr., Wood, T. A., & Merbler, J. B. (1980). An orientation and mobility programme for multiply impaired blind children. *Exceptional Children, 46*, 326–331.

Harris, H. (1980). *The principles of human biochemical genetics*, 3rd edition. Amsterdam: Elsevier/ North Holland Biomedical Press.

Harris, P. L. (1989). *Children and emotions*. Blackwell, Oxford.

Harris, P. L., Brown, E., Marriot, C., Whittall, S., & Harmer, S. (1991). Monsters, ghosts and witches: Testing the limits of the fantasy reality distinction in young children. *British Journal of Developmental Psychology, 9*, 105–123.

Hatwell, Y. (2003). Intermodal coordination in children and adults. In Y. Hatwell, A. Streri, & E. Gentaz (Eds.), *Touching for knowing. Cognitive psychology of tactile manual perception.* (pp. 207–219). Amsterdam: John Benjamins Publishing Company.

Hatwell, Y. (1994). Cross-modal transfer and cross-modal integration. In M. Richelle, J. Requin, & M. Roberts (Eds.), *Hand book of experimental psychology* (pp. 543–584). Paris: Presses Universitaires de France.

Hatwell, Y. (1987). Motor and cognitive functions of the hand in infancy and childhood. *International Journal of Behavioural Development, 20*, 509–526.

Hatwell, Y., Streri, A., & Gentaz, E. (2003). *Touching for knowing. Cognitive psychology of tactile manual perception*. Amsterdam: John Benjamins Publishing Company.

Hatta, T. (1978). The functional asymmetry of tactile pattern learning in normal subjects. *Psychologia: An International Journal of Psychology in the Orient, 21*(2), 83–89.

Hatta, T., & Koike, M. (1991). Left hand preference in frightened mother monkeys in taking up their babies. *Neuropsychologia, 29*, 207–209.

Hauser, M. D., & Anderson, K. (1994). Left hemisphere dominance for processing vocalizations in adults, but not infant rhesus monkeys.: Field experiments. *Proceedings of the National Academy of Sciences, USA, 91*, 3946–3948.

Hauser, M. D., & Schon Ybarra, M. (1994). The role of lip configuration in monkey vocalizations: Experiments using xylocaine as a nerve block. *Brain and Language, 46*, 232–244.

Hauser, M. D., Evans, C. S., & Marler, P. (1993). The role of articulation in the production of rhesus monkey, Macaca mulatta, vocalizations. *Animal Behaviour, 45*, 423–433.

Hausman, M., Kirk, I. J., & Corballis, M. C. (2004). Influence of task complexity on manual symmetries. *Cortex, 40*, 103–110.

Haxby, J. V., Gobbini, M. I., Furey, M. L., Ishai, A., & Schouten, J. L., et al. (2001). Distributed and overlapping representations of faces and objects inventral temporal cortex. *Science, 293*, 2425–2430.

Hay, L. (1990). Developmental changes in eye hand coordination behaviours: Preprogramming versus feedback control. In C. Bard, M. Fleury, & L. Hay (Eds.), *Development of eye hand coordination across the life span* (pp. 217–244). Columbia, SC: University of South Carolina Press.

Hay, L. (1978). Accuracy of children on an open-loop pointing task. *Perceptual Motor Skills, 47*, 1079–1082.

Hay, L., Bard, C., Fleury, M., & Teasdale, N. (1991). Kinematics of aiming in direction and amplitude: A developmental study. *Acta Psychologica, 77*, 203–215.

Head, H., & Holmes, G. (1911). Sensory disturbances from cerebral lesions. *Brain, 34*, 102–254.

Healey, J. M., Lederman, J., & Geschwind, N. (1986). Handedness is not an unidimensional trait. *Cortex, 22*, 33–53.

Heffner, H. E., & Heffner, R. S. (1986). Effects of unilateral and bilateral auditory cortex lesions on the discrimination of vocalization by Japanese Macques. *Journal of Neurophysiology, 56*, 683–701.

Hollins, M. (1986). Mental haptic rotation: More consistent in blind subjects. *Journal of Visual Impairment and Blindness, 80*, 950–952.

Hecaen, H., & Albert, M. L. (1978). *Human neuropsychology*. New York: Wiley.

Hecaen, H., & Piercy, M. (1956). Paroxysmal dysphasia and the problem of cerebral dominance. *Journal of Neurology, Neurosurgery and Psychiatry, 18*, 194–201.

Hecaen, H., & Sauguet, J. (1971). Cerebral dominance in left handed subjects. *Cortex, 7*, 19–48.

Hecaen, H., DeAgostini, M., & Monzon-Montes, A. (1981). Cerebral organization in left handers. *Brain and Language, 12*, 261–284.

Heilbroner, P. L., & Holloway, R. L. (1988). Anatomical brain asymmetries in New World and Old World monkeys: Stages of temporal lobe development in primate evolution. *American Journal of Physical Anthropology, 16*, 38–48.

Hein, A., & Diamond, R. M. (1971). Contrasting development of visually triggered and guided movements in kittens with respect to interocular and interline equivalence. *Journal of Comparative and Physiological Psychology, 76*, 219–224.

Hein, A., Held, R., & Gower, E. R. (1970). Development and segmentation of visually controlled movement by selective exposure during rearing. *Journal of Comparative and Physiological Psychology, 73*, 181–187.

Held, R. (1965). Plasticity in sensory motor systems. *Scientific American, 213*, 84–94.

Held, R. (1963). Plasticity in human sensory motor control. *Science, 142,* 455–462.

Held, R., & Bauer, J. A. (1967). Visualy guided reaching in infant monkeys after restricted rearing. *Science, 155,* 718–720.

Held, R., & Hein, A. (1967). On the modifiability of form perception. In W. Wathen-Dunn (Ed.), *Models for the perception of speech and visual form* (pp. 296–304). Cambridge, Massachussets: MIT Press.

Held, R., & Hein, A. (1963). A movement produced stimulation in the development of visually guided behaviour. *Journal of Comparative and Physiological Psychology, 56,* 872–876.

Held, R., & Hein, A. (1958). Adaptation of disarranged hand-eye coordination contingent upon re-afferent stimulation. *Perceptual and Motor Skills, 8,* 87–90.

Hepp-Reymond, M–C., Heusler, E. J., & Maier, M. A. (1996). Precision grip in humans: Temporal and spatial synergies. In A. M. Wing, P. Haggard, & J. R. Flanagan (Eds.), *The Neuropsychology and physiology of hand movements* (pp. 37–68). San Diego: Academic Press.

Hermelin, B., & O'Connor, N. (1971). Functional asymmetry in the reading of Braille. *Neuropsychologia, 9,* 431–435.

Herrmann, D. J., & Van Dyke, K. (1978). Handedness and the mental rotation of perceived patterns. *Cortex, 14,* 521–529.

Hershenson, M. (1964). Visual discrimination in the human newborn. *Journal of Comparative and Physiological Psychology, 43,* 112–122.

Hobson, P. R. (1991). Against the theory of mind. *British Journal of Developmental Psychology, 9,* 33–51.

von Hofsten, C. (1991). Structuring of early reaching movements: A longitudinal study. *Journal of Motor Behaviour, 23,* 280–292.

von Hofsten, C. (1990). A perception-action perspective on the development of manual movements. In M. Jeannerod (Ed.), *Attention and performance, XIII. Motor representation and control* (pp. 739–762). Hillsdale, NJ: Erlbaum.

von Hofsten, C., & Ronnqvist, L. (1988). Preparation for grasping an object: A developmental study. *Journal of Experimental Psychology: Human Perception and Performance, 4,* 610–621.

von Hofsten, C., & Rosblad, B. (1988). The integration of sensory information in the development of precise manual pointing. *Neuropsychologia, 26,* 805–821.

von Hofsten, C. (1982). Eye hand coordination in the newborn. *Developmental Psychology, 18,* 450–461.

Hollins, M. (1986). Mental haptic rotation: More consistent in blind subjects. *Journal of Visual Impairment and Blindness, 80,* 950–952.

Holloway, R. L., & de la Coste-Lareymondie, M. C. (1982). Brain endocast asymmetry in pongids and hominids. Some preliminary findings on the paleontology of cerebral dominance. *American Journal of Physical Anthropology, 58,* 101–110.

Holmes, N. P., Calvert, G., & Spence, C. (2007). Tool use changes multisensory task interactions in seconds: Evidence from the cross-modal congruency task. *Experimental Brain Research, 183,* 465–475.

Holmes, N. P., Calvert, G. A., & Spence, C. (2004). Extending or projecting peripersonal space with tools? Multisensory interactions highlight only the distal and proximal ends of tools. *Neuroscience Letters, 372,* 62–67.

von Holst, E., & Mittelstaedt, H. (1950). Das reaffenzprinzip (wechselwirkengen zeischenzentral Nervensystem und peripherie). *Naturwissenschaften, 37,* 464–476.

Honda, H. (1984). Functional between- hand differences and out flow eye position information. *Quarterly Journal of Experimental Psychology, 36 A,* 75–88.

Honda, H. (1982). Rightward superiority of eye movements in a bimanual aiming task. *Quarterly Journal of Experimental Psychology, 34A*, 449–513.

Hopkins, B. (1993). *On the developmental origins of handedness. Annual report. Research and clinical centre for child development.* Sapporo, Japan: Hokkaido University.

Hopkins, W. D. (1997). Hemispheric specialization for local and global processing of hierarchical visual stimuli. In chimpanzees (Pan troglodytes). *Neuropsychologia, 35*, 343–348.

Hopkins, W. D., & Leavens, D. A. (1998). Hand use and gestural communication in chimpanzees (Pan troglodytes). *Journal of Comparative Psychology, 112*, 95–99.

Hopkins, W. D., & Bard, K. A. (1995). Asymmetries in spontaneous head orientation in infant chimpanzees (Pan trologdytes). *Behavioural Neuroscience, 109*, 808–812.

Horn, G. (1985). *Memory, imprinting and the brain.* Oxford: Clarendon Press.

Horn, G., Bradley, P., & Bateson, P. P. G. (1979). An audiographic study of the chick brain after imprinting. *Brain Research, 168*, 361–373.

Howard, I. P., Craske, B., & Templeton, W. B. (1965). Visuomotor adaptation to discordant exafferent stimulation. *Journal of Experimental Psychology, 70*, 189–191.

Hugdahl, K. (1995). Classical conditioning and implicit learning: The right hemisphere hypothesis. In R. J. Davidson, & K. Hugdahl (Eds.), *Brain asymmetry* (pp. 235–267). Cambridge, MA: MIT Press.

Hyvarinen, J., Hyvärinen, L., & Linnankoski, I. (1991). Modification of parietal association cortex and functional blindness after binocular deprivation in young monkeys. *Experimental Brain Research, 42*, 1–8.

Hyvarinen, J., Carlson, S., & Hyvarinen, L. (1981). Early visual deprivation alters modality of neuronal responses in area 19 of the monkey cortex. *Neuroscience Letters, 26*, 239.

Hyvarinen, J., & Poranen, A. (1974). Function of the parietal associative area 7 as revealed from cellular discharges in alert monkeys. *Brain, 97*, 673–692.

Hume, D. (1739/1960). *Treatise of human nature.* Oxford: Oxford University Press.

Ittyerah, M. (2009). Hand ability and practice in congenitally blind children. *Journal of Development and Physical Disabilities, 21*, 329–344.

Ittyerah, M. (2000). Hand skill and hand preference in blind and sighted children. *Laterality: Asymmetries of Body, Brain, and Cognition, 5*(3), 221–235.

Ittyerah, M. (1996). Do the hands differ in skill? *Brain and Cognition, 32*(2), 291–296.

Ittyerah, M. (1993). Hand preferences and hand ability in congenitally blind children. *Quarterly Journal of Experimental Psychology, 46B*, 35–50.

Ittyerah, M., & Marks, L. E. (2008). Intra-modal and cross-modal discrimination of curvature: Haptic touch versus vision. *Current Letters in Psychology, 24*(1), 1–11.

Ittyerah, M., & Gaunet, F., Rossetti, Y. (2007). Pointing with the left and right hands in congenitally blind children. *Brain and Cognition, 64*(2), 170–183.

Ittyerah, M., & Marks, L. E. (2007). Perception and memory in curvature stimuli. Haptic touch versus vision. *British Journal of Psychology, 98*, 589–610.

Ittyerah, M., & Sharma, R. (1997). The performance of hearing impaired children in handedness and perceptual motor tasks. *The Genetic, Social and General Psychology Monographs, 123*(3), 285–302.

Ittyerah, M., & Goyal, M. (1997). Fantasy and reality distinction of congenitally blind child- ren. *Perceptual and Motor Skill, 85*, 897–898.

Ittyerah, M., & Samarapungavan, A. (1989). The performance of congenitally blind children in cognitive developmental tasks. *British Journal of Developmental Psychology, 7*, 129–139.

Ittyerah, M. (1984). Verbal and modality specific coding in inter and intra-modal matching among children and adults. *Perceptual and Motor Skills, 59*, 797–798.

Ittyerah, M., & Broota, K. D. (1983). Inter and intra-modal processing of sensory specific stimuli. *Perceptual and Motor skills, 56*, 507–517.

Iriki, A., Tanaka, M., Obayashi, S., & Iwamura, Y. (2001). Selfimages in the video monitor coded by monkey intraparietal neurons. *Neuroscience Research, 40*, 163–173.

Iriki, A., Tanaka, M., & Iwamura, Y. (1996). Coding of modified body schema during tool use by macaque postcentral neurones. *Neuroreport, 7*, 2325–2330.

Jacquet, A–Y., Esseily, R., Rider, D., & Fagard, J. (2012). Handedness for grasping objects and declarative pointing: A longitudinal study. *Developmental Psychobiology, 54*, 36–46.

Ja'gr, J. (1991). Two ways of integrating the time factor in complex motor patterns. *Homeostatis in health and disease, 33*, 135–136.

Jakobson, L. S., Archibald, Y. M., Carey, D. P., & Goodlae, M. A. (1991). A kinematic analysis of reaching and grasping movements in a patient recovering from optic ataxia. *Neuropsychologia, 29*, 803–809.

James, T. W., Humphrey, G. K., Gati, J. S., Servos, P., Menon, R. S., & Goodale, M. A. (2002). Haptic study of threedimensional objects activates extrastriate visual areas. *Neuropsychologia* 40, 1706–1714.

Jastrow, J. (1889). The perception of space by different senses. *Mind, 11*, 538–544.

Jay, R. (1986). *Learned pigs and fireproof women*. New York: Warner Books.

Jeannerod, M. (1997a). *The cognitive neuroscience of action*. Cambridge, Mass: Blackwell Publishers.

Jeannerod, M. (1997b). Grasping objects: The hand as a pattern recognition device. In M. C. Hepp-Reymond, & G. Marini (Eds.), *Perspectives of motor behaviour and its neural basis* (pp. 19–32). Basel: Karger.

Jeannerod, M. (1994). The hand and the object: The role of the posterior parietal cortex in forming motor representations. *Canadian Journal of Physiology and Pharmacology, 72*, 535–541.

Jeannerod, M. (1988). *The neural and behavioural organization of goal directed movements*. Oxford: Clarendon Press.

Jeanneord, M. (1984). The timing of neural prehension movements. *Journal of Motor Behaviour, 16*, 235–264.

Jeannerod, M. (1981). Intersegmental coordination during reaching at natural visual objects. In A. Baddeley (Ed.), *Attention and Performance*, Vol. 9 (pp. 153–168). Hillsdale: Erlbaum.

Jeannerod, M., & Rossetti, Y. (1993). Visuo motor coordination as a dissociable function: Experimental and clinical evidence. In C. Kennard (Ed.), *Visual perceptual defects* (pp. 439–460). Bailliere's clinical Neurology, International practice and research, London: Bailliere, Tindall.

Jeannerod, M., & Decety, J. (1990). The accuracy of visuo-motor transformation. An investigation into the mechanisms of visual recognition of objects. In M. Goodale (Ed.), *Vision and action: The control of grasping* (pp. 33–48). Norwood: Ablex.

Jefferies, R. P. S. (1979). Calcichordates. In R. W. Fairbridge & D. Jablonski (Eds.), *Encyclopedia of earth sciences. vol. 7: The Encyclopedia of paleontology* (pp. 161–167). Strondsburg, PA: Dowden, Hutchinson & Ross Inc.

Jefferies, R. P. S., & Lewis, D. N. (1978). The English Silurian Fossil Polacocystites forebesianus and the ancestors of the vertebrates. *Philisophical Transations of the Royal Society of London B, 282*, 204–223.

Jerrison, H. (1973). *Evolution of the brain and intelligence*. New York: Academic Press.

Johansson, R. S., & Lamotte, R. H. (1983). Tactile detection thresholds for a single asperity on an otherwise smooth surface. *Somatosensory and Motor Research, 1*, 21–31.

Johansson, R. S., Westling, G., Baeckstrom, A., & Flanagan, J. R. (2001). Eye hand coordination in object manipulation. *Journal of Neuroscience, 21*, 6917–6932.

Johansson, R. S., & Westling, G. (1988a). Coordinated isometric muscle commands adequately and errorneously programmed for the weight during lifting task with the precision grip. *Experimental brain research, 66*, 141–154.

Johansson, R. S., & Westling, G. (1988b). Programmed and triggered actions to rapid load changes during precision grip. *Experimental Brain Resarch, 66*, 141–154.

Johansson, R. S., & Westling, G. (1984). Roles of glabrous skin receptors and sensory motor memory in automatic control of precision grip when lifting rougher or more slippery objects. *Experimental Brain Research, 56*, 550–564.

Johnson, S. P., & Aslin, R. N. (1995). Perception of object unity in 2-month-old infants. *Developmental Psychology, 31*, 739–745.

Johnston, A. N. B., & Rose, S. P. R. (2002). Memory and lateralized recall. In L. J. Rogers, & R. J. Andrew (Eds.), *Comparative vertebrate lateralization* (pp. 533–581). New York: Cambridge University Press.

Jones, S. A., Cressman, E. K., & Henriques, D. Y. (2010). Proprioceptive localization of the left and right hands. *Experimental Brain Research, 204*(3), 373–383.

Kabbash, P., Mackenzie, I. S., & Buxton, W. (1993). Human performance using computer input devices in the preferred and non preferred hands, Proceedings of Inter CHI 1993, 474–481.

Kanizsa, G., Renzi, P., Conte, S., Compostela, C., & Guerani, L. (1993). Amodal completion in mouse vision. *Perception, 22*, 713–722.

Kappers, A. M. L. (2007). Haptic space processing. Allocentric an egocentric reference frames. *Canadian Journal of Experimental Psychology, 61*, 208–218.

Kappers, A. M. L. (2005). Intermediate frames of reference in haptically perceived parallelity. In A. Bicchi, & M. Bergamasco (Eds.), *World Haptics Conference (WHC 2005): First joint euro-haptics conference and symposia on haptic interfaces for virtual environment and teleoperater systems* (pp. 3–11). Los Alamitos, CA: IEEE Computer Society.

Kappers, A. M. L. (2004). The contributions of egocentric and allocentric reference frames in haptic spatial tasks. *Acta Psychologia, 117*, 333–340.

Kappers, A. M. L. (2003). Large system deviations in a bimanual parallelity task: Further analysis of contributing factors. *Acta Psychologia, 114*, 131–145.

Kappers, A. M. L. (2002). Haptic perception of parallelity in the midsagittal plane. *Acta Psychologia, 109*, 25–40.

Kappers, A. M. L., & Koenderink, J. J. (1996). Haptic unilateral and bilateral discrimination of curved surfaces. *Perception, 25*, 739–749.

Katz, D. (1989). *The world of touch*, edited and translated by L. Krueger. Hillsdale, NJ: Erlbaum (originally published, 1925, as Der Aufbau der Tastwelt).

Kaufman, E., Lord, M., Reese, T., & Volkman, J. (1949). The discrimination of visual number. *American Journal of Psychology, 62*, 498–525.

Kellman, P. J., & Spelke, E. S. (1983). Perception of partly occluded objects in infancy. *Cognitive Psychology, 15*, 483–524.

Keller, H., Katner, J. Borke, J., Yovis, R., & Kleis, A. (2005). Parenting styles and the development of the categorical self: A longitudinal study on mirrir self-recognition in Cameroonian Nso and German families. *International Journal of Behavioiural Development, 29*, 496–504.

Keller, H., Yovis, R., Borke, J., Kartner, J., & Jensen, H., et al. (2004). Developmental consequences of early parenting experiences: Self recognition nd self regulation in three cultural communities. *Child Development, 6,* 1745–1760.

Kennett, S., Taylor, C., & Haggard, P. (2001). Non informative vision improves spatial resolution of touch in humans. *Current Biology, 11,* 1188–1191.

Kertesz, A., & Geshwind, N. (1971). Patterns of pyramidal decussation and their relationship to handedness. *Archives of Neurology, 24,* 326–332.

Khan, A. Z., Pisella, L., Vighetto, A., Cotton, F., Luaute, J., Boisson, D., et al. (2005). Optic ataxia errors depend on remapped, not viewed, target location. *Nature Neuroscience, 8*(4), 418–420.

King, J. E., & Landau, V. I. (1993). Manual preferences in varietiesof reaching in squirrel monkeys. In J. P. Ward, & W. D. Hopkins (Ed.), *Primate laterality: Current evidence of primate asymmetries* (pp. 107–124). New York: Springer- Verlag.

Kiphart, M. J., Hughes, J. L., Simmons, J. P., & Cross, H. A. (1992). Short term haptic memory for complex objects. *Bulletin of the Psychonomic Society, 30,* 212–214.

Kohlberg, L. (1969). Stage and sequence: The cognitive developmental approach to socialization. In D. A. Goslin (Eds.), *Handbook of socialization, theory and research* (pp. 347–480). Chicago: Rand McNally.

Kohler, I. (1964). The formation and transformation of the perceptual world. Psychological Issues, (Whole No.4).

Komai, T., & Fukuaka, G. (1934). A study on the frequency of left handedness and left footedness among Japanese school children. *Human Biology, 6,* 33–42.

Kimura, D. (1993). *Neuromotor mechanisms in human communication.* New York: Oxford University Press.

Kimura, D. (1982). Left hemisphere control of oral and brachial movements and their relation to communication. *Philosophical Transactions of the Royal Society, London, 292*(1089), 135–149.

Kimura, D. (1961). Cerebral dominance and the perception of verbal stimuli. *Canadian Journal of Psychology, 15,* 166–171.

Krauthamer, G. (1968). Form perception across sensory modalities. *Neuropsychologia, 6,* 105–113.

Kumar, S. (1977). Short term memory for a nonverbal tactual task after cerebral commisurrotomy. *Cortex, 13,* 55–61.

Kumari, L. (1976). *Conservation abilities of six to eight year old boys from five occupat- ional environments.* Unpublished Master's dissertation, University of Delhi.

Kupers, R., Pietrini, P., Ricciardi, E., & Ptito, M. (2011). The nature of consciousness in the visually deprived brain. *Frontiers in Psychology, 2,* doi: 10.3389/fpsyg.2011.00019.

Kuypers, H. G. (1982). A new look at the organization of the motor system. *Progress in Brain research, 57,* 381–404.

Laabs, G. J. (1973). Retention characteristics of different reproduction cues in motor short-term memory. *Journal of Experimental Psychology, 100,* 168–177.

Laabs, G. J., & Simmons, R. W. (1981). Motor memory. In D. Holding (Ed.), *Human skills* (pp. 119–151). New York: Wiley.

Lacreuse, A., & Fragaszy, D. M. (1999). Left hand preferences in capuchins (Cebus paella): Role of spatial demands in manual activity. *Laterality, 4,* 65–78.

Lakatos, S., & Marks, L. E. (1999). Haptic form perception: Relative salience of local and global features. *Perception and Psychophysics, 61,* 895–908.

LaMendola, N. P., & Bever, T. G. (1997). Peripheral and cerebral asymmetries in the rat. *Science, 278*, 483–486.

Landau, B., Spelke, E., & Gleitman, H., (1984). Spatial knowledge in a young blind child. *Cognition, 16*, 225-260.

Landau, B., Gleitman, H., & Spelke, E. (1981). Spatial knowledge and geometric representation in a child blind from birth. *Science, 213*, 1275-1278.

Landsmeer, J. M. F. (1962). 'Power grip and precision handling.' *Annals of Rheumatic Diseases, 21*, 164–170.

Landsmeer, J. M. F., & Long, C. (1965). The mechanisms of finger control based on electromyograms and location analysis. *Acta Anatomica, 60*, 330–347.

Langdon, D., & Warrinton, E. K. The role of the left hemisphere in spatial processing. *Cortex, 36*, 691–702 (2000).

Laurendeau, M., & Pinard, A. (1962). *Causal thinking in the child.* New York: International University Press.

Lea, S. E. G. (1984). *Instinct, environment and behaviour.* London: Metheun.

Leary, R. F., & Allendoff, F. W. (1989). Fluctuating asymmetry as an indicator of stress: Implication for conservation biology. *Trends in Ecology and Evolution, 4*, 214–217.

Leask, S. J., & Crow, T. J. (1997). How far does the brain lateralize? An unbiased method for determining the optimum degree of hemisphere specialization. *Neuropsychologia, 35*, 1381–1387.

Lebron-Rodreguez, D. E., & Pasnak, R. (1977). Introduction of the intellectual gains in blind children. *Journal of Experimental Child Psychology, 24*, 505–515.

Lechelt, E. C. (1982). The stability of tactile laterality differences across manipulations of temporal and spatial stimulus patterning. *International Journal of Rehabilitation Research, 5*, 370–372.

Lee, D. N., & Aronson, E. (1974). Visual proprioceptive control of standing in human infants. *Perception & Psychophysics, 15*, 529–532.

Lee, D. N., Young, D. S., Reddish, P. E., Lough, S., & Clayton, T. M. H. (1983). Visual timing in hitting an accelerating ball. *Quarterly Journal of Experimental Psychology, 35A*, 333–346.

Lederman S. J., & Klatzky R. L. (1987). Hand movements: A window into haptic object recognition. *Cognitive Psychology, 19*, 342–368.

Leinonen, L., Hyvarinen, J., Nyman, G., & Linankoski, I. (1979). Functional properties of neurons in lateral part of associative area 7 in awake monkeys. *Experimental Brain Research, 34*, 299–320.

Leonard, J. A., & Newman, R. C. (1967). Spatial orientation in the blind. *Nature, 215*(5108), 1413–1414.

Lloyd, D. M., Shore, D. I., Spence, C. & Calvert, G. A. (2003). Multisensory representation of limb position in human premotor cortex. *Nature Neuroscience 6*, 17–18.

LeMay, M. (1977). Asymmetries of the skull and handedness. *Phrenology revisited. Journal of the Neurological Sciences, 32*, 243–253.

LeMay, M., & Geshwind, N. (1975). Hemispheric differences in the brains of great apes. *Brain, Behaviour and Evolution, 11*, 48–52.

Levander, M., & Schalling, D. (1988). Hand preference in a population of Swedish college students. *Cortex, 24*, 149–156.

Levy, J., & Trevarthen, C. (1977). Perceptual, semantic and phonetic aspects of elementary language processes in split brain patients. *Brain, 100*, 105–118.

Lewis, J. W., 2006. Cortical networks related to human use of tools. *The Neuroscientist, 12*(3), 211–231.

Liben, L. S. (1988). Conceptual issues in the development of spatial cognition. In J. Stiles-Davis, M. Kritchevsky, & U. Bellugi (Eds.), *Spatial cognition* (pp. 167–194). Hillsdale, New Jersey: LEA.

Lobb, H. (1965). Vision versus touch in form discrimination. *Canadian Journal of Psychology, 19*, 175–186.

Lock, A. (Ed.). 1978. *Action, gesture and symbol: The emergence of language*. London: Academic Press.

Locke, J. (1689/1975). *An essay concerning human understanding*. Oxford: Oxford University Press.

Lockman, J. J., & Ashmead, D. H. (1983). Asynchronies in the development of manual behaviour. In L. P. Lipsitt (Ed.), *Advances in infancy research*, Vol. 2 (pp. 113–136). Norwood, NJ: Ablex.

Logie, R. H. (1986). Visuo-spatial processing in working memory. *Quarterly Journal of Experimental Psychology, 38A*, 229–247.

Loomis, J. M., & Lederman, S. J. (1986). Tactual perception. In K. R. Boff, L. Kaufman, & J. P. Thomas (Eds.), *Handbook of perception and human performance. Vol. II. Cognitive processes and performance* (pp. 31(1)–31(41)). New York: John Wiley & Sons, Inc.

Lonton, A. P. (1976). Hand preference in children with myelomeningocele and hydrocephalus. *Developmental Medicine and Child Neurology, 18*(Suppl. 37), 143–149.

Lorincz, E., & Fabre-Thrope, M. (1996). Shift of laterality and compared analysis of paw performances in cats during practice of a visuomotor task. *Journal of Comparative Psychology, 110*, 307–315.

Louw, S. (2002). *Haptic discrimination of geometrical object properties*. Unpublished doctoral dissertation, Helmholtz Instituut, University of Utrecht, Netherlands.

Lowe P. A., Mayfield J. W., & Reynolds, L. R. (2003). Gender differences in memory test Performances among children and adolescents. *Archives of Clinical Neurology, 18*, 865–878.

Lynch, A., Lee, H. M., Bhat, M., & Galloway, J. C. (2008). No stable arm preference during pre- reaching period: A comparison of right and left hand kinematics with and without a toy present. *Developmental Psychobiology, 50*, 390–398.

Mackenzie, C. L., & Iberall, T. (1994). *The grasping hand*. Amsterdam: North Holland.

Macaluso, E., Frith, C. D., & Driver, J. (2002). Crossmodal spatial influences of touch on extrastriate visual areas take current gaze direction into account. *Neuron, 34*, 647–658.

McCabe, B. J. (1991). Hemispheric asymmetry of learning-induced changes. In J. P. Ewert, R. R. Capranica, & D. J. Ingle (Eds.), *Advances in vertebrate neuroethology* (pp. 262–276). New York: Plenum.

McClelland, J. C., Rumelhart, D. E., & Group PDP Research Group. (1986). *Parallel distributed processing: Explorations in the microstructure of cognition. Vol. 2: Psychological and biological models*. MIT press, Cambridge, MA.

MacNeilage, P. F. (1998). Towards a unified view of cerebral hemispheric specializations in vertebrates. In A. Milner (Ed.), *Comparative neuropsychology* (pp. 167–183). Oxford: Oxford University Press.

MacNeilage, P. F., Studdert-Kennedy, M. G., & Lindbolm, B. (1987). Primate handedness reconsidered. *Behavioural and Brian Sciences, 10*, 247–303.

Macmillan, N., & Creelman, C. D. (1991). *Signal detection theory: A user's manual*. Cambridge: Cambridge University Press.

Manning, J. T., & Chamberlain, A. T. (1990). The left side cradling preference in great apes. *Animal Behaviour, 39*, 1224–1227.

Manoel E. J. & Connolly K. J. (1997). Variability and stability in the development of skilled actions. In K. J. Connolly, & H. Forssberg (Eds.), *The neurophysiology and neuropsychology of motor development. Clinics in developmental medicine, No 143/144* (pp. 286–318). London: Mac Keith Press.

McCall, G. N., & Cunningham, N. M. (1971). Two point discrimination: Asymmetry in spatial discrimination on the two sides of the tongue, a preliminary report. *Perceptual and Motor Skills, 32*, 368–370.

McDonnell, P. M. (1979). Patterns of eye hand co ordination in the first year of life. *Canadian Journal of Psychology, 33*, 253–267.

McGonigle, D. J., Hanninen, R., Salenius, S., Hari, R., Frackowiak, R. S. J., & Frith, C. D., 2002. Whose arm is it anyway? An fMRI case study of supernumerary phantom limb. *Brain, 125*, 1265–1274.

McGraw, M. B. (1945). *The neuromuscular maturation of the human infant.* New York: Columbia University Press.

McGrew, W. C., & Marchant, L. F. (1999). Laterality of hand use pays off in foraging success for wild chimpanzees. *Primates, 40*, 509–513.

Mahrer, P., & Miles, C. (2002). Recognition for tactile sequences. Memory, *10*, 7–20.

McManus I. C. (1991). The inheritance of left handedness. In G. R. Bock, & J. Marsh (Eds.), *Biological asymmetry and handedness*, Ciba Foundation Symposium 162. New York, USA: John Wiley & Sons Inc.

McManus, I. C., Suk, G., Cole, D. R., Mellon, A. F., Wong, J. & Kloss, J. (1988). The development of handedness in children. *British Journal of Developmental Psychology, 6*, 257–272.

McManus, I. C. (1985). Handedness, language dominance and aphasia: A genetic model. *Psychological Medicine*, Monograph Supplement, No.8.

McNutt, J. C. (1975). Asymmetry in two point discrimination on the tongues of adults and children. *Journal of Communication disorders, 8*, 212–320.

Mandler, G., & Shebo, B. J. (1982). Subitizing: An analysis of its component processes. *Journal of Experimental psychology: General, 111*, 1–22.

Maravita, A., & Iriki, A., (2004). Tools for the body schema. *Trends in Cognitive Sciences, 8*, 79–86.

Maravita, A., Spence, C., & Driver, J. (2003). Multisensory integration and the body schema: Close to hand and within reach. *Current Biology, 13*, R531–R539.

Maravita, A., Clarke, K., Husain, M., & Driver, J. (2002). Active tool use with contralesional hand can reduce crossmodal extinction oftouch on that hand. *Neurocase, 8*, 411–416.

Maravita, A., Spence, C., Kennett, S., & Driver, J. (2002). Tool use changes multimodal spatial interactions between vision and touch in normal humans. *Cognition, 83*, B24–B34.

Marks, L. E. (1978). *Unity of the senses. Interrelations between the modalities.* New York: Academic Press.

Marr, D. (1982). *Vision: A computational investigation into the human representation and processing of visual information.* San Francisco: W. H. Freeman.

Marteniuk, R. G., Leavitt, J. L., MacKenzie, C. L., & Athenes, S. (1990). Functional relationships between grasp and transport components in a prehension task. *Human Movement Science, 9*, 149–176.

Marzke, M. (1992). Evolutionary development of the human thumb. *Hand Clinics, 8*, 1–8.

Marzke, M., & Wullstein, K. (1996). Chimpanzee and human grips: A new classification with a focus on evolutionary morphology. *International Journal of Primatology, 17*, 117–139.

Matoba, M., Masataka, N., & Tanioka, Y. (1991). Cross-generational continuity of hand-use preferences in marmosets. *Behaviour, 117*, 281–286.

McCarty, M. E., Clifton, R. K., Ashmead, D. H., Lee, P., & Goubet, N. (2001). How infants use vision for grasping objects. *Child Development, 72*(4), 973–987.

McIntyre, J., Stratta, F., & Lacquaniti, F. (1998). Short-term memory for reaching to visual targets: Psychophysical evidence for body-centered reference frames. *The Journal of Neuroscience, 18*(20), 8423–8435.

Mead, M. (1932). An investigation of the thought of the primitive children, with special reference to animism. *Journal of the Royal Anthropological Institute of Great Britain and Ireland, 62*, 173–190.

Medendorp, W. P., Goltz, H. C., Vilis, T., & Crawford, J. D. (2003). Gazecentered updating of visual space in human parietal cortex. *Journal of Neuroscience, 23*(15), 6209–6214.

Merabet, L. B., & Pascual-Leone, A. (2010). Neural reorganization following sensory loss: The opportunity of change. *Nature Reviews Neuroscience, 11*, 44–52.

Meunier, H., Vauclair, J., & Fagard, J. (2012). Human infants and baboons show the same pattern of handedness for a communicative gesture. *PLoS ONE*, 7(3), e33959.

Meyer, V., & Yates, A. J. (1955). Intellectual changes following temporal lobectomy for psychomotor epilepsy. *Journal of Neurology, Neurosurgery and Psychiatry, 18*, 44–52.

McClelland, J., Rumelhart, J., & Group, P. R. (1980). Parallel distributed processing: Explanations in the microstructure of cognition. *Psychological and Biological Models, 2*, Cambridge, MA: MIT Press.

McKenzie, R. (1991). *Lateralization of auditory and visual learning in the domestic chick.* D. Phil thesis, University of Sussex.

McNeill, D. (1992). *Hand and mind: What gestures reveal about thought.* Chiago :University of Chicago Press.

Michael, G. F. (1981). Right handedness. A consequence of infant supine head orientation preference. *Science, 212*, 685–687.

Michel, G. F. (1988). A lateral bias in the neuropsychological functioning of human infants. *Developmental Neuropsychology, 14*, 45–69.

Miklosi, A., & Andrew, R, J (1999). Right eye use associated with decision to bite in zebrafish. *Behavioural Brain Research, 105*, 199–205.

Millar, S. (2008). *Space and Sense.* Hove and New York: Psychology Press.

Millar, S., & Al-Attar, Z. (2005). What aspects of vision facilitate haptic processing? *Brain and Cognition, 59*, 258–268.

Millar, S., & Al-Attar, Z. (2004). External and body-centered frames of reference in spatial memory: Evidence from touch. *Perception & Psychophysics, 66*(1), 51–59.

Millar, S., & Al-Attar, Z. (2003a). How do people remember spatial information from tactile maps? *British Journal of Visual Impairment, 21*, 64–72.

Millar, S., & Al-Attar, Z. (2003b). Spatial reference and scanning with the left and right hand. *Perception, 32*, 1499–1511.

Millar, S. (1999a). Memory in touch. *Psicothema, 11*, 746–767.

Millar, S. (1997). *Reading by touch.* London: Routledge.

Millar, S. (1994). *Understanding and representing space. Theory and evidence from studies with blind and sighted children.* Oxford: Clarendon Press.

Millar, S. (1991). A reverse lag in the recognition and production of tactual drawings: Theoretical implications for haptic coding. In M. Heller, & W. Schiff (Ed.), *The psychology of touch* (pp. 301–325). Hillsdale, New Jersey: Lawrence Erlbaum Associates.

Millar, S. (1988). Models of sensory deprivation: The nature/nurture dichotomy and spatial representation in the blind. *International Journal of Behavioural Development, 11*, 69–87.

Millar, S. (1987a). The perceptual window in two handed braille. *Do the left and right hands process braille simultaneously? Cortex, 23*, 111–122.

Millar, S. (1987b). Perceptual and task factors in fluent Braille. *Perception, 16*, 521–536.

Millar, S. (1986a). Aspects of size, shape and texture in touch: Redundancy and interference in children's discrimination of raised dot patterns. *Journal of Child Psychology and Psychiatry, 27*, 367–381.

Millar, S. (1986b). Drawing as image and representation in blind children. In D. G. Russel, D. Marks, & T. E. Richardson (Eds.), *Image 2* (pp. 254–259). Dunedin, New Zealand: Human Performance Associates.

Millar, S. (1985b). Movement cues and body orientation in recall of locations of blind and sighted children. *Quarterly Journal of Experimental Psychology, 37A*, 257–279.

Millar, S. (1981a). Crossmodal and intersensory perception and the blind. In R. D. Walk, & H. L. Jr. Pick (Eds.), *Intersensory perception and sensory integration* (pp. 281–314). New York: Plenum Press.

Millar, S. (1981b). Self- referent and movement cues in coding spatial location by blind and sighted children. *Perception, 10*, 255–264.

Millar, S. (1978). Short-term serial tactual recall: Effects of grouping tactually probed recall of Braille letters and nonsense shapes by blind children. *British Journal of Psychology, 69*, 17–24.

Millar, S. (1977a). Early stages of tactual matching. *Perception, 6*, 333–343.

Millar, S. (1977b). Tactual and name matching by blind children. *British Journal of Psychology, 68*, 377–387.

Millar, S. (1976). Spatial representation by blind and sighted children. *Journal of Experimental Child Psychology, 21*, 460–479.

Millar, S. (1975b). Effects of tactual and phonological similarity on the recall of Braille letters by blind children. *British Journal of Psychology, 66*, 193–201.

Millar, S. (1975c). Effects of phonological and tactual similarity on serial object recall by blind and sighted children. *Cortex, 11*, 170–180.

Millar, S. (1971). Visual and haptic cue utilization by preschool children. The recognition of visual and haptic stimuli presented separately and together. *Journal of Experimental Child Psychology, 12*, 88–94.

Millar, S., & Ittyerah, M. (1991). Movement imagery in young and congenitally blind children: Mental practice without visuo-spatial information. *International Journal of Behavioral Development, 15*, 125–146.

Milner, A. D., Dijkerman, H. C., McIntosh, R. D. Y., Rossetti, Y., & Pisella, L. (2003). Delayed reaching and grasping in patients with optic ataxia. *Progress in Brain Research, 142*, 225–242.

Milner, B. (1982). Some cognitive effects of frontal lobe lesions in man. *Philosophical transactions of the Royal Society of London, B298*, 211–226.

Milner, B. (1974). Hemispheric specialization: Scope and limits. In F. O. Schmitt, & G. G. Warden (Eds.), *The neurosciences, third study programme* (pp. 75–89). Cambridge, Mass: MIT Press.

Milner, B. (1962). Laterality effects in audition. In V. B. Mountcastle (Ed.), *Interhemispheric relations and cerebral dominance* (pp. 177–195). Baltimore: John Hopkins.

Milner, B., & Taylor, L. (1972). Right hemisphere superiority in tactile pattern recognition after cerebral commissurotomy. *Evidence of non verbal memory. Neuropsychologia, 10*, 1–17.

Milner, B., Branch, C., & Rasmussen, T. (1964). Observations on cerebral dominance. In A. V. S. De Reuck, & M. O'connor (Eds.), *Disorders of language* (CIBA Foundation Symposium). London: Churchill.

Milner D. A., & Goodale, M. A. (1993). Visual pathways to perception and action. In T. P., Hicks, S. Molotchnikoff, & T. Ono (Eds.), *Progress in brain research*, Vol. 95 (pp. 317–337). Amsterdam: Elsevier.

Mishkin, M., Ungerleider, L. G., & Macko, K. M. (1983). Object vision and spatial vision: Two cortical pathways. *Trends in Neuroscience*, 6, 414–417.

Moore, C., & Furrow, D. (1991). The development of the language of belief. The expression of relative certainty. In D. Frye, & C. Moore (Eds.), *Children's theories of mind. Mental states and social understanding*. Hillsdale, NJ, Erlbaum.

Moore, K. C. (1896). The mental development of a child. *Psychological Reviews, Monographs Supplement, 1*(3).

Morris, R. D., & Hopkins, W. D. (1993). Perception of human chimeric faces by chimpanzees: Evidence for a right hemisphere advantage. *Brain and Cognition, 21*, 111–122.

Mosquera, M., Geribas, N., Bagallo, A., Llorente, M., & Riba, D. (2012). Complex tasks force hand laterality and technological behavior in naturalistically housed chimpanzees: Inferences in human evolution. *The Scientific World Journal, 2012*, doi: 10. 1100/2012/514809.

Mountcastle, V. B. (2005). *The sensory hand. neural mechanisms of somatic sensation*. Massachusetts, and London, England: Harvard University Press Cambridge.

Mikalelian, H., & Held, R. (1964). *Two types of adaptation to an optically rotated visual field. American Journal of Psychology, 77*, 257–263.

Murata, K., Cramer, H., & Bach-y-Rita, P. (1965). Neuronal convergence of noxious, acoustic and visual stimuli in the visual cortex of the cat. *Journal of Neurophysiology, 28*, 1223–1239.

Nachshon, I., & Carmon, A. (1975). Hand preference in sequential and spatial discrimination tasks. *Cortex, 11*, 123–131.

Napier, J. (1956). The prehensile movements of the human hand. *Journal of bone and joint surgery, 38-B*(4), 902–913.

Napier, J. (1970). *The roots of mankind*. London: Allen &Unwin.

Nathan, P. W., Smith, M. C., & Deacon, P. (1990). The spinocortical tracts in man. *Brain, 113*, 303–324.

Natsopoulos, D., Kiosseoglou, G., & Xeromeritou, A. (1992). Handedness and spatial ability in children: Further support for Geshwind's hypothesis pathology of superiority and for Annett's theory of intelligence. *Genetic, Social and General Psychology Monographs, 118*, 103–126.

Navon, D. (1977). Forest before trees. The precedence of global features in visual perception. *Cognitive Psychology, 9*, 353–383.

Nebes, R. D. (1971). Superiority of the minor hemisphere in commissurotomized man for the perception of part-whole relations. *Cortex, 7*, 333–349.

Nebes, R. D. (1972). Dominance of the minor hemisphere in commissurotomized man on a test of figural unification. *Brain, 95*, 633–638.

Nefs, H. (2002). *Tactual discrimination ofgratings*. Unpublished doctoral dissertation, Helmholtz Instituut, University of Utrecht, Netherlands.

Nefs, H. T., Kappers, A. M. L., & Koenderink, J. J. (2001). Amplitude and spatial period discrimination in sinusoidal gratings by dynamic touch. *Perception, 30*, 1263–1274.

Neville, H. J. (1988). Cerebral organization for spatial attention. In J. Stiles-Davies, M. Kntechevsky, & U. Bellugi (Eds.), *Spatial cognition: Brain bases and development* (pp. 327–341). Hillsdale, NJ: LEA.

Newcombe, N., & Huttenlocher, J. (2000). *Making space: The development of spatial representation and reasoning.* Cambridge: MIT Press.

Newell, F. N. (2004). Crossmodal object recognition. In C. Spence, G. Calvert, & B. Stein (Eds.) *Handbook of multisensory integration* (pp. 123–140). Cambridge, MA: MIT Press.

Newell, K. M., & Cesari, P. (1998). Body scale and the development of hand form and function in prehension. In K. J. Connolly (Ed.), *The psychobiology of the hand* (pp. 232–256). MacKeith Press; Clinics in Developmental Medicine, No.47.

Newell, K. M. (1986). Constraints on the development of coordination. In M. G. Wade, H. T. A. Whiting (Eds.), *Motor development in children: Aspects of coordination and control* (pp. 341–360). Boston: Martinus Nijhoff.

Newport, R., Rabb, B., & Jackson, S. R. (2002). Non informative vision improves haptic spatial perception. *Current Biology, 12*, 1661–1664.

Nilsson, L.-G., Nyberg, L., Klingberg, T., Aberg, C., & Persson, J., et al. (2000). Activity in motor areas while remembering action events. *Neuroreport, 11*, 2199–2201.

Norman, J. F., Norman, H. F., Clayton, A. M., Lianekhammy, J., & Zielke, G. (2004). The visual and haptic perception of natural object shape. *Perception & Psychophysics, 66*, 342–357.

Norman, J. F., Dawson, T. E., & Raines, S. R. (2000). The perception and recognition of natural object shape from deforming and static shadows. *Perception, 29*, 135–148.

Nottebohm, F. (1980). Brain pathways for vocal learning in birds: A review of the first ten years In J. M. Sprague, & A. N. Epstein (Eds.), *Progress in psychobiology and physiological psychology* (pp. 85–124). New York: Academic Press.

Nottebohm, F. (1977). Asymmetries in neural control of vocalization in the canary. In S. Harnard, R. W. Doty, L. Goldstein, J. Jaynes, & G. Krauthamer, (Eds.) *Lateralization of the nervous system*, (pp. 23–44). New York: Academic Press.

Nottebohm, F. (1971). Neural lateralization of vocal control in a Passerine bird. I. Song. *Journal of Experimental Zoology, 177*, 229–261.

Nottebohm, F. (1970). Ontogeny of bird song. *Science, 176*, 950–956.

Oeztuerk, C., Durmazlar, N., Ural, B., Karaagaoglu, E., Yalaz, K., & Anlar, B. (1999). Hand and eye preference in normal preschool children. *Clinical Pediatrics, 38*, 677–680.

Olson, GM Sherman , T. (1983). Attention, learning, and memory in infants. In P. H. Mussen (Ed.), *Handbook of child psychology: Vol. 2. Infancy and developmental psychobiology* (pp. 1001–1080). New York: Wiley.

Orton, S. T. (1937). *Reading, writing and speech problems in children*. New York: Norton.

Osada, Y., & Schiller, P. H. (1994). Can monkeys see objects under conditions of transparency and occlusion? *Investigative Ophalthalmology and Visual Science, 35*, 1664.

Oscar-Berman, M., Rehbein, L., Porfert, A., & Goodglass, H. (1978). Dichhaptic hand order effects with verbal and nonverbal tactile stimulation. *Brain and Language, 6*, 323–333.

Ostrovsky, Y., Andalman, A., & Sinha, P. (2006). Acquisition of visual function after extended congenital blindness. *Psychological Science, 17*, 1009–1014.

Paillard, J. (1991). Motor and representational framing of space. In J. Palliard (Ed.), *Brain and space* (pp. 63–181). Oxford: Oxford University Press.

Paillard, J. (1987). Cognitive vs. sensorimotor encoding of spatial information. In P. Ellen, & C. Thinus-Blanc (Eds.), *Cognitive processes and spatial orientation in animal and man* (pp. 43–77). Dordrecht: Martinus Nijhoff.

Palliard, J., Michel, F., & Stelmach, G. (1983). Localization without content. A tactile analog of 'Blindsight". *Archives of Neurology, 40*, 548–551.

Palmer, R. D. (1964). Development of a differentiated handedness. *Psychological Bulletin, 62*, 257–272.

Parson, T. D., Larson, P., Kratz, K., & Thiebeux, N. (2004). Sex differences in mental rotation and spatial rotation in a virtual environment. *Neuropsychologia, 42*, 555–562.

Pasini, M., & Tessari, A. (2001). Hemispheric specialization in quantification processes. *Psychological Research, 65*, 57–63.

Penfield, W., & Rasmussen, T. (1950). *The cerebral cortex of man*. New York: MacMillan.

Perenin, M. T. (1991). Discrimination of motion direction in perimetrically blind fields. *Neuroreport, 2*, 397–400.

Perenin, M. T., & Jeannerod, M. (1975). Residual vision in cortically blind hemifields. *Neuropsychologia, 13*, 1–7.

Perenin, M. T., & Vighetto, A. (1998). Optic ataxia: A specific disruption in visuomotor mechanisms. Different aspects of the deficit in reaching in objects. *Brain, 111*, 643–674.

Perenin, M. T., & Rossetti, Y. (1996). Grasping without form discrimination in an hemianopic field. *Neuroreport, 7*, 793–797.

Peters, M. (1991a). Sex, handedness mathematical ability. *Canadian journal of Psychology, 45*, 415–419.

Peters, M. (1991b). *Laterality and motor control In Petitto, L. A. (1983). From gesture to symbol: The relationship between form and meaning in the acquisition of personal pronouns in American Sign Language.* Unpublished Doctoral Dissertation, Harvard University, Cambridge, MA.

Peters, M. (1990a). Interaction of vocal and manual movements. In G. Hammond (Ed.), *Cerebral control of speech and limb movement. Advances in psychology* (pp. 535–574). North Holland.

Peters, M. (1990b). Subclassification of left handers poses problems for theories handedness. *Neuropsychologia, 28*, 279–289.

Peters, M. (1990c). Phenotype in normal lefthanders. An understanding of phenotype is the basis for understanding mechanism and inheritance of handedness. In S. Coren (Ed.), *Left handedness. Behavioural implications and anomalies* (pp. 167–192). North Holland: Elsevier Science Publishers.

Peters, M. (1987). A nontrivial motor performance difference between right handers and left handers: Attention as intervening variable in the expression of handedness. *Canadian Journal of Psychology, 41*, 91–99.

Peters, M. (1985). Constraints in the coordination of bimanual movements and their expression in skilled and unskilled subjects. *Quaterly Journal of Experimental Psychology, A37*, 171–196.

Peters, M. (1983). Differentiation and lateral specialization in motor development. In G. Young, S. J. Segalowitz, C. M. Cortex, & S. E. Trehub (Eds.), *Manual specialization and the developing brain* (pp. 141–159). New York: Academic Press.

Peters, M. (1980). Why the preferred hand taps more quickly than the non preferred hand. Three experiments on handedness. *Canadian Journal Psychology, 34*, 62–71.

Peters, M. (1976). Prolonged practice of a simple motor task by preferred and non preferred hands. *Perceptual and Motor skills, 43*, 447–450.

Peters, M., & Durding, B. (1978). Handedness measured by finger tapping. A continuos variable. *Canadian Journal of Psychology, 32*, 257–261.

Peterson, L. R., & Peterson, M. J. (1959). Short-termretention of individual verbal items. *Journal of Experimental Psychology, 58*, 193–198.

Petersen, M. R., Beecher, M. D., Zoloth, S. R., Green, S., & Marler, P. R., et al. (1984). Neural lateralization of vocalizations by Japanese Macques. Communicative significance is more important than acoustic structure. *Behavioural Neuroscience, 98*, 779–790.

Petersen, M. R., Beecher, M. D., Zoloth, S. R., Moody, D. B., & Stebbins, W. C. (1978). Neural laterlization of species-specific vocalizations in Japanese macaques (Macaca fuscata). *Science, 202*, 324–327.

Petitto, L. A. (1991). Language ontogeny: Evidence from studies of sign language. Paper presented to the Stanford Child Language Forum.

Piaget, J. (1954). *The construction of reality in the child.* New York: basic Books.

Piaget, J. (1953). *Origin of intelligence in the child.* London: Routeledge and Kegan Paul.

Piaget, J. (1929). *The child's conception of the world.* London: Routledge & Kegan Paul.

Piazza, M., Mechelli, A., Price, C. J., & Butterworth, B. (2006). Exact and approximate judgements of visual and auditory numerosity: An fMRI study. *Brain Research, 1106*, 177–188.

Pieraut-le- Bonniec, G. (1985). Hand eye coordination and infant's construction of convexity and concavity. *British Journal of Developmental Psychology, 3*, 273–280.

Pietrini, P., Ptito, M., & Kupers, R. (2009). "Blindness and consciousness consciousness: New lights from the dark". In S. Laureys, & G. Tononi (Eds.), *The neurology of consciousness* (pp. 360–374). New York: Academic Press.

Pietrini, P., Furey, M. L., Ricciardi, E., Gobbini, M. I., & Wu, W. H., et al. (2004). Beyond sensory images: Object-based representation in the human ventral pathway. *Proceedings of the National Academy of Sciences of the United States of America, 101*, 5658–5663.

Pillow, B. H. (1989). Early understanding of perception as a source of knowledge. *Journal Experimental Child Psychology, 47*, 116–129.

Pineau, A., & Streri, A. (1990). Intermodal transfer of spatial arrangement of the component parts of an object in 4/5 month-old- infants. *Perception, 19*, 795–804.

Plaisier, M. A., Bergmann Tiest, W. M., & Kappers, A. M. L. (2010a). Grabbing subitizing with both hands- Bimanual number processing. *Experimental Brain Research*, DOI 10.1007/s00221-009-2146-1.

Plaisier, M. A., Bergmann Tiest, W. M., & Kappers, A. M. L. (2010b). Haptic object individuation. IEEE Transactions on Haptics.

Plaisier, M. A., Bergmann Tiest, W. M., & Kappers, A. M. L. (2009b). One, two, three, many-subitizing in active touch. *Acta Psychologia, 131*, 163–170.

Plaisier, M. A., Bergmann Tiest, W. M., & Kappers, A. M. L. (2009a). Salient features in three dimensional haptic space perception. *Attention, Perception and Psychophysics, 71*, 421–430.

Plaisier, M. A., Bergmann Tiest, W. M., & Kappers, A. M. L. (2008a). Haptic pop out in a hand sweep. *Acta Psychologia, 128*, 368–377.

Plato, C. C., Fox, K. M., & Garruto, R. M. (1984). Measures of lateral functional dominance: Hand dominance. *Human Biology, 56*, 259–276.

Pont, S., Kappers, A. M. L., & Koenderink, J. J. (1999). Similar mechanisms underlie curvature comparison by static and dynamic touch. *Perception & Psychophysics, 61*, 874–894.

Pont, S., Kappers, A. M. L., & Koenderink, J. J. (1997). Haptic curvature discrimination at several regions of the hand. *Perception and Psychophysics, 59*, 1225–1240.

Pont, S., Kappers, A. M. L., & Koenderink, J. J. (1998). The influence of stimulus tilt on haptic curvature matching and discrimination by dynamic touch. *Perception, 27*, 869–880.

Porac, C., & Coren, S. (1981). *Lateral preferences and human behaviour*. New York: Springer-Verlag.

Porac, C., Coren, S., & Duncan, P. (1980). Life span age trends in laterality. *Journal of Gerontology, 35*, 715–721.

Posner, M. J., & Petersen, S. E. (1990). The attention system of the human brain. *Annual Review of Neuroscience, 13*, 25–42.

Posner, M. I., Boies, S. J., Eichelman, W. H., & Taylor, R. L. 1969. Retention of visual and name codes of single letters. *Journal of Experimental Psychology, 79*(1, Pt. 2), 1–16.

Postma, A., Zuidhoek, S., Noordzij, M. L., Kappers, A. M. (2007). Differences between early-blind, late-blind and blindfolded-sighted people in haptic spatial-configuration learning and resulting memory traces. *Perception, 36*(8), 1253–1265.

Postman, A., Jager, G., Kensels, R. P., & Koppenschaar, H. P. (2004). Sex differences for selective forms of spatial memory. *Brain and Cognition, 54*, 24–34.

Poston, B., Enoka, J. A., Enoka, R. M. (2008). Practice and endpoint accuracy with the left and right hands of old adults: The right-hemisphere aging model. *Muscle Nerve, 37*(3), 376–386.

Pratt, C., & Bryant, P. E. (1990). Young children understand that looking leads to knowing so long as they are looking through a single barrel. *Child Development, 61*, 973–982.

Provins, K. A. (1997). Handedness and speech: A critical reappraisal of the role of genetic and environmental factors in the cerebral lateralization of function. *Psychological Review, 104*, 554–571.

Pryde, K. M., Bryden, P. J., & Roy, E. A. (2000). A developmental analysis of the relationship between hand preference and performance: I. Preferential reaching into hemispace. *Brain and Cognition, 43*, 370–374.

Ptito, M., & Desgent, S. (2006). Sensory input-based adaptation and brain architecture. In P. Baltes, P. Lorenz, & F. Rösler (Eds.), *Lifespan development and the brain* (pp. 111–123). Cambridge, UK: Cambridge University Press.

Ptito, M., Moesgaard, S., Gjedde, A., & Kupers, R. (2005). Cross-modal plasticity revealed by electrotactile stimulation of the tongue in the congenitally blind. *Brain* 128, 606–614.

Putnam, H. (1987). *The many faces of realism*. Cambridge, England: Cambridge University Press.

Ramsay, D. S. (1980). Beginnigs of bimanual handedness and speech in infants. *Infant Behaviour and Development, 3*, 67–77.

Ramsay, D. S. (1979). Manual preference for tapping infants. *Developmental Psychology, 15*, 437–442.

Rashid, N. Y., & Andrew, R. J. (1989). Right hemisphere advantage for tographical orientation in the domestic chick. *Neuropsychologia, 27*, 937–948.

Rasmussen, T., & Milner, B. (1977). The role of early left brain injury in determining lateralization of cerebral speech functions. *Annals of the New York Academy of Sciences, 299*, 355–369.

Rasmussen, T., & Penfield, W. (1947). Further studies of the sensory and motor cerebral cortex of man. *Federation Proceedings, Federation of American Societies for Experimental Biology*, 6, 452–460.

Rebillard, G., Carlier, E., Rebillard, M., & Pujol, R. (1977). Enhancement of visual responses on the primary auditory cortex of the cat after an early destruction of cochlear receptors. *Brain Research, 129*, 162–164.

Reed, C. L., Caselli, R. J., & Farah, M. J. (1996). Tactile agnosia: Underlying impairment and implications for normal tactile object recognition. *Brain, 119*, 875–888.

Renier, L. A., Anurova, I., De Volder, A. G., Carlson, S., & VanMeter, J., et al. (2010). Preserved functional specialization for spatial processing in the middle occipital gyrus of the early blind. *Neuron, 68*, 138–148.

Révész, G. (1950). *Psychology and art of the blind.* New York: Longman, Green.

Riesen, A. H., & Aarons, L. (1959). Visual movement and intensity discrimination in cats after early deprivation of pattern vision. *Journal of Comparative and Physiological Psychology, 52*, 142–149.

Riggio, L., Delantionio, A., & Barato, M. Right hand superiority in serial recall of tactile stimuli. *Cortex, 18*, 595–601 (1980).

Rizzolatti, G., Matelli, M., & Pavesi, G. (1983). Deficits in attention and movement following the removal of postarcuate (area 6) and prearcuate (area 8) cortex in macaque monkeys. *Brain, 106*, 655–673.

Robins, A., Lippolis, G., Bisazza, A., Vallortigara, G., & Rogers, L. J. (1998). Lateralized of agnostic responses and hind- limb use in toads. *Animal Behaviour, 56*, 875–881.

Robinson, D. L., Goldberg, M. E., & Stanton, G. B. (1978). Parietal association cortex in the primate: Sensory mechanisms and behaviour modulation. *Journal of Neurophysiology, 41*, 910–932.

Roberts, J. R., & Bauman, M. K. (1944). *Minnesota rate of manipulation test. Pennsylvania Bimanual work sample—adapted for the blind.* Minneapolis: Educational Publishers, Inc.

Rock, I., & Victor, J. (1964). Vision and touch: An experimentally created conflict between two senses. *Science, 143*, 594–596.

Rochat, P. (1985). From hand to mouth and eye: Development of intermodal exploration by young infants. Paper presented at the eighth meeting of the International Society for the study of Behavioural Development. Tours: France.

Roeder, M. B., Mahone, M. E. J., Gidley Larson, J. G., Mostofsky, S. H., & Cutting, L. E., et al. (2008). Left-right differences on timed motor examination in children. *Child Neuro Psychology, 14*, 249–262.

Roessner, V., Banaschewski, T., Uebel, H., Becker, A., & Rothenberger, A. (2004). Neuronal network models of ADHD – Lateralization with respect to interhemispheric connectivity reconsidered. *European Child & Adolescent Psychiatry, 13*, 71–79.

Rogers, L. J. (2002). Advantages and disadvantages of lateralization. In L. S. Rogers, & R. J. Andrew (Ed.), *Comparative vertebrate lateralization* (pp. 126–153). New York: Cambridge University Press.

Rogers, L. J., & Andrew, R. J. (2002). *Comparative vertebrate lateralization.* New York: Cambridge University Press.

Rogers, L. J. (2000). Evolution of hemispheric specialization: Advantages and disadvantages. *Brain and Language, 73*, 236–253.

Rogers, L. J. (1997). Early experimental effects on laterality Research on chicks has relevance to other species. *Laterality, 2*, 199–219.

Rogers, L. J. (1991). Development of lateralization. In R. J. Andres (Ed.), *Neural and behavioural plasticity: The use of the domestic chick as a model*, (pp. 507–535). Oxford: Oxford University Press.

Rogers, L. J. (1990). Light input and the reversal of functional lateralization in the chicken brain. *Behavioural Brain Research, 38*, 211–221.

Rogers, L. J. (1989). Laterality in animals. *International Journal of Comparative Psychology, 3*, 5–25.

Rogers, L. J. (1982). Light experience and asymmetry of brain function in chickens. *Nature, 297*, 223–225.

Rogers, L. J. (1980). Lateralization in the avain bird. *Bird Behaviour, 2*, 1–12.

Rogers, L. J., Andrew, R. J., & Burne, T. H. J. (1999). Light exposure of the embryo and development of behavioiural lateralization in chicks: I. Olfactory responses. *Behavioural Brain research, 97*, 195–200.

Rogers, L. J., & Kaplan, G. (1996). Hand preferences and other lateral biases in rehabilitated orang-utans (Pongo pygmaeus pygmaeus). *Animal Behaviour, 51*, 13–25.

Rogers, L. J., & Workman, L. (1989). Light exposure during incubation affects competitive behavior in domestic chicks. *Applied Animal Behaviour Science, 23*, 187–198.

Rogers, L. J., & Anson, J. M. (1979). Lateralization of function in the chicken forebrain. *Pharmacology, Biochemestry and Behaviour, 10*, 679–686.

Rosenbaum, D A., Meulenbrock, R. J., & Vaughan, J (1999). Remembered positions: Stored locations or stored postures? *Experimental Brain Research, 124*, 503–512.

Rossetti, Y. (1998). Implicit short-lived motor representations of space in brain damaged and healthy subjects. *Consciousness and Cognition, 7*, 520–558.

Rossetti, Y., & Pisella, L. (2002). Several vision for action systems: A guide to dissociating an disintegrating dorsal and ventral functions. In W. Prinz, & B. Hommel (Eds.), *Attention and performance XIX; Common mechanisms in perception and action* (pp. 62–119). Oxford: Oxford University Press.

Rossetti, Y., Pisella, L., & Pelisson, D. (2000). Eye blindness and hand sight: Temporal aspects of visuomotor processing. *Visual Cognition, 7*, 785–808.

Rossetti, Y., Rode, G., & Boisson, D. (1988). Implicit processing of somaesthetic information: A dissociation between where and how? *Neuroreport, 6*, 505–510.

Rossetti, Y., Gaunet, F., & Thinus-Blanc, C. (1996). Early visual experience affects memorization and spatial representation of proprioceptive targets. *Neuroreport, 7*, 1219–1223.

Rossetti, Y., & Regnier, C. (1995). Representation in action: Pointing to a target with various representations. In B. G., Bardy, R. J., Bootsma, & Y. Guiard (Eds.), *Studies in perception and action* (pp. 233–236). Mahwah, N. J.: Lawrence Erlbaum Associates, Inc.

Rossetti, Y., Regnier, C., Perenin, M. T., Rode, G., Lacquaniti, F., & Boisson, D. (1995). Actions et representations: Influence de la memorization et de la verbalization du but sur les mouvements de patients et les sujets sains. Annual meeting of the French Neuroscience association (Societe des Neurosciences), Lyon, France.

Rossi, G. F., & Rosadini, G. (1967). Experimental analysis of cerebral dominance in man. In F. L. Darley (Ed.), *Brain mechanisms underlying speech and language* (pp. 167–174). New York: Grune & Stratton.

Roy, E. A., Bryden, P., & Cavill, S. (2003). Hand differences in pegboard performance through development. *Brain and Cognition, 53*, 315–317.

Rudel, R. G., Denckla, M. B., & Hirsch, S. (1977). The development of left hand Superiority for discriminating Braille configurations. *Neurology, 27*, 160–164.

Rudel, R. G., & Teuber, H. L. (1964). Crossmodal transfer of shape information by children. *Neuropsychologia, 2*, 1–8.

Rushworth, M. F. S., Nixon, P. D., Renowden, S., Wade, D. T., & Passingham, R. E. (1997). The left parietal cortex and attention to action. *Neuropsychologia, 35*, 1261–1273.

Sadato, N., Pascual-Leone, A., Grafman, J., Ibanez, V., & Deiber, M. P., et al. (1996). Activation of the primary visual cortex by Braille reading in blind subjects. *Nature, 380*, 526–528.

Sakata, H., & Taira, M. (1994). Parietal control of hand action. *Current opinion in Neurobiology, 4*, 847–856.

Salapatek, P., & Kessen, W. (1966). Visual scanning of triangles by the human newborn. *Journal of Experimental Child Psychology, 15*, 22–29.

Saling, M. M., & Kaplan-Solms, K. (1989). On lateral asymmetry in cradling behavior and breast sensitivity. *Current Anthropology, 30*, 210–211.

Satz, P. (1980). Incidence of aphasia in left handers: A test of some hypothetical models of cerebral speech organization. In J. Herron (Ed.), *Neuropsychology of left handedness* (pp. 189–198). New York: Academic Press.

Salk, L. (1973). The role of the heartbeat in the relations between mother and infant. *Scientific American, 228*, 24–29.

Salk, L. (1962). Mother's heartbeat as an imprinting stimulus. *Transactions of the New York Academy of Sciences, 24*, 753–763.

Salk, L. (1960). The effects of the normal heartbeat sound on the behavior of the new-born infant: Implication for mental health. *World Mental Health, 12*, 168–175.

Savelsberg, G. J. P., & van der Kamp, J. (1994). The effect of body orientation to gravity on early infant reaching. *Journal of Experimental Child Psychology, 58*, 510–528.

Schaefer, M., Rothemund, Y., Heinze, H.-J., & Rotte, M. (2004). Short-term plasticity of the primary somatosensory cortex during tool use. *NeuroReport 15*(8), 1293–1297.

Schaffer, H. R. (Ed.). (1977). *Studies in mother-infant interaction*. London: Academic Press.

Schieber, M. H. (1995). Muscular production of individuated finger movements: The roles of extrinsic finger muscles. *Journal of Neuroscience, 15*, 284–297.

Schluter, N. D., Krams, M., Rushworth, M. F. S., & Passingham, R. E. (2001). Cerebral dominance for action in the human brain. *Neuropsychologia, 39*, 105–113.

Schneider, G. E. (1969). Two visual systems. *Science, 163*, 895–902.

Schulman-Galambos, C., & Galambos, R. (1979). Brainstem evoked response audiometry in newborn hearing screening. *Archives of Otolaryngology, 105*, 86–90.

Schutz, R. W., & Roy, E. A. (1973). Absolute error: The devil in disguise. *Journal of Motor Behavior, 5*, 141–153.

Schwartz, A. B., Moran, D. W., & Reina, G. A. (2004). Differential representation of perception and action in the frontal cortex. *Science, 303*(5656), 380–383.

Semjen, A., Summers, J. J., & Cattaert, D. (1995). Hand coordination in bimanual circle drawing. *Journal of Experimental Psychology: Human Perception and Performance, 21*, 1139–1157.

Semmes, J. (1968). Hemispheric specialzation: A possible clue to mechanism. *Neuropsychologia, 6*, 11–26.

Von Sendon, M. (1960). *Space and sight: The perception of space and shape in congenitally blind patients before and after operation*. London: Methuen.

Sergent, J., & Signoret, J. L. (1992). Functional and anatomical decomposition of face processing: Evidence from prosopagnosia and PET study of normal subjects. *Philosophical Transactions of the Royal Society of London B, 335*, 55–62.

Sharma, R. (1998). A *neuro- developmental profile of hearing impaired children*. Unpublished doctoral dissertation. University of Delhi, Delhi 110007.

Sherrington, C. (1906). *The integrative action of the nervous system*. New York: Scribner.

Sherwood, D. E. (1994). Hand preferences, practice order and spatial assimilation in rapid bimanual movement. *Journal of Motor Behaviour, 26*, 123–134.

Shen, L. M., & Alexander, G. E. (1997b). Preferential representation of instructed target location versus limb trajectory in dorsal premotor area. *Journal of Neurophysiology, 77*(3), 1195–1212.

Sherrick, C. E., & Craig, J. C. (1982). The psychophysics of touch. In W. Schiff, & E. Foulke (Ed.), *Tacual perception: A sourcebook* (pp. 55–81). New York: American Foundation for the Blind.

Shinn, M. W. (1893). Notes on the development of a child. *University of California Public Education, 1*, 178–236.

Shirley, M. M. (1931). *The first two years: A study of twenty-five babies*, Vol. 1. Locomotor Development, Minneapolis: University of Minnesota Press.

Smith, A. (1966). Intellectual functions in patients with lateralized frontal tumour. *Journal of Neurology, Neurosurgery and Psychiatry, 29*, 52–59.

Smith, M. O., Chu, J., & Edmonston, W. E. (1977). Cerebral lateralization of haptic perception: Interaction of responses to Braille and music reveals a functional basis. *Science, 197*, 689–690.

Smyth, M. M., Katamba, J., & Peacock, K. A. (2004). Development of prehension between 5 and 10 years of age: Distance scaling, grip aperture, and sight of the hand. *Journal of Motor Behavior, 36*, 91–103.

Smyth, M. M., Peacock, K. A., & Katamba, J. (1994). The role of sight of the hand in the development of prehension in childhood. *The Quarterly Journal of Experimental Psychology, 57A*, 269–296.

Soechting, J. F., Engel, K. C., & Flanders, M. (2001). The Duncker illusion and eye hand coordination. *The Journal of Neurophysiology, 85*, 843–854.

Soper, H. V., Satz, P., Orsini, D. L., Henry, R. R., Zvi, J. C., & Schulman, M. C. (1986). Handedness patterns in autism suggest subtypes. *Journal of Autism and Developmental Disorders, 16*, 155–167.

Sovrano, V. A., Bisazza, A., & Vallortigara, G. (2001). Lateralization of response to social stimuli in fishes: A comparison between different methods and species. *Physiology and Behaviour, 74*, 237–244.

Sperry, R. W. (1974). Lateral specialization in the surgically separated hemispheres. In F. O. Schmidt, & F. G. Worden (Eds.), *The neurosciences: Third study program* (pp. 5–19). Cambridge, Mass: MIT Press.

Spiegler, B. J., & Yeni Komshian, G. H. (1983). Incidence of left handed writing in a college population with reference to family patterns of hand preference. *Neuropsychologia, 21*, 651–659.

Steinhuis, R. E., & Bryden, M. P. (1989). Different dimensions of hand preference that relate to skill and unskilled activities. *Cortex, 25*, 289–304.

Steinhuis, R. E., & Bryden, M. P. (1990). Reliability of hand preference items and factors. *Journal of Clinical and Experimental Neuropsychology, 12*(6), 921–930.

Stelmach, G. E., & Larish, D. D. (1980). Egocentric referents in human limb orientation. In G. E. Stelmach, & J. Requin (Ed.), *Tutorials in motor behaviour* (pp. 167–184). Amsterdam: North-Holland Publishing Co.

Stocket, T. W. & Weigelt, M. (2012). Plasticity of human handedness: Decreased one hand bias and intermanual performance asymmetry in expert basket ball players. *Journal of Sports Science, 30*, 1037–1045.

Streri, A. (2005). Touching for knowing in infancy: The development of manual abilities in very young infants. *European Journal of Developmental Psychology, 4*, 325–343.

Streri, A. (2002). Hand preference in 4-month-old infants: Global or local processing of objects in the haptic mode. *Current Psychology Letters, 7*, 39–50.

Streri, A., & Gentaz, D. (2003). Crossmodal recognition of shape from hand to eyes in newborns. *Sensory and Motor Research, 20*, 13–18.

Streri, A., Lhote, M., & Dutilleul, S. (2000). Haptic perception in newborns. *Developmental Science, 3*, 319–327.

Streri, A., & Gouarir, C. (1996). Handedness: Left–right differences in object holding and motor skills in 6-month-old infants. *Current Psychology on Cognition, 15*, 209–230.

Streri, A., & Molina, M. (1993). Visual-tactual and Tactual-visual transfer between objects and pictures in 2-month-old infants. *Perception, 22*, 1299–1318.

Streri, A. (1993). *Seeing, reaching, touching. The relations between vision and touch in infancy.* London: Harvester, Wheatsheaf.

Streri, A., & Milhet, S. (1988). Equivalences intermodales de la forme des objets entre la vision et le toucher chez les bébés de 2 mois. *L'Année Psychologique, 88*, 329–341.

Streri, A. (1987). Tactile discrimination of shape and intermodal transfer in 2- to 3- month old infants. *British Journal of Developmental Psychology, 5*, 213–220.

Streri, A., & Pêcheux, M. G. (1986a). Crossmodal transfer of form in five month old infants. *British Journal of Developmental Psychology, 4*, 161–167.

Streri, A., & Pêcheux, M. G. (1986b). Tactual habituation and discrimination of form in infancy: A comparison with vision. *Child Development, 57*, 100–104.

Stevens, S. S., & Guirao, M. (1963). Subjective scaling of length and area and matching of length to loudness and brightness. *Journal of Experimental Psychology, 66*, 177–186.

Stein, J. F. (1991). Space and the parietal association areas. In J. Palliard (Ed.), *Brain and space* (pp. 185–222). Oxford: Oxford University Press.

Steinfield, G. J. (1967). Concepts of set and availability and their relation to the reorganization of ambiguous pictorial stimuli. *Psychological Review, 74*, 505–522.

Stuphorn, V., Bauswein, E., & Hoffman, K.-P. (2000). Neurons in the primate superior colliculus coding for arm movements in gaze-related coordinates. *Journal of Neurophysiology, 83*, 1283–1299.

Spelke, E. S. (1981). The infants acquisition of knowledge of bimodally specified events. *Journal of Experimental Child Psychology, 31*, 279–299.

Spelke, E. S. (1979). Perceiving bimodally specified events in infancy. *Developmental Psychology, 15*, 626–636.

Subramanian, K. K. (2006). (Ed.). *The handicrafts industry in Kerala. Blending heritage with economics. Studies in local development* 3. Delhi: Dannish Books.

Sugarman-Bell, S. (1978). Some organizational aspects of preverbal communication. In I. Markova (Ed.), *The social context of language* (pp. 49–66). New York: Wiley.

Sullivan, E. V., & Turvey, M. T. (1972). Short term retention of tactile stimulation. *Quarterly Journal of Experimental Psychology, 24*, 253–261.

Tan, U. (1990). The left brain determines the degree of left handedness. *International Journal of Neuroscience, 53*, 75–85.

Tan, L. E. (1985). Laterality and motor skills in four year olds. *Child Development, 56*, 119–124.

Tapley, S. M., & Bryden M. P. (1985). A group test for the assessment of performance between the hands. *Neuropsychologia, 23*, 215–221.

Taub, E., & Berman, A. J. (1968). Movement and learning in the absence of sensory feedback. In X. Homewood (Ed.), *The neuropsychology of spatially oriented behaviour* (pp. 173–192). Illinois: Dorsey Press.

Teghtsoonian, M. (1965). The judgment of size. *American Journal of Psychology, 36*, 392–402.

Teng, E. L., Lee, P. H., Yang, K. S., & Chang, P. C. (1976). Handedness in a Chinese population. *Biological, social and pathological factors. Science, 19*, 1148–1150.

Theurel, A., Frileux, S., Hatwell, Y., & Gentaz, E. (2012). The haptic recognition of geometrical shapes in congenitally blind and blindfolded adolescents: Is there a haptic prototype effect? *Plos One, 2147*(6), e40251.

Tobin, M. J., & Hunter, B. (1974). Assessing the manual dexterity of the visually handicapped. *New Beacon, VIII*, 169–172.

Todor, J. I., Doane, T. (1978). Handedness and hemispheric asymmetry in the control of Movements. *Journal of Motor behaviour, 10*, 295–300.

Toldi, J., Rojik, I., & Fehér, O. (1994). Neonatal monocular enucleation induced cross-modal effects observedin the cortex of adult rat. *Neuroscience, 62*, 105–114.

Tomasello, M. (1993). On the interpersonal origins of self concept. In U. Neisser (Ed.), *The perceived self: Ecological and interpersonal sources of self knowledge* (pp. 174–184). New York: Cambridge University Press.

Tommasi, L., & Vallortigara, G. (1998). Footedness in binocular and monocular chicks. *Laterality, 4*, 89–95.

Tommasi, L., Vallortigara, G., & Zanforlin, M. (1997). Young chicks learn to localize the centre of a spatial environment. *Journal of Comparative Physiology A, 180*, 567–572.

Touwen, B. C. L. (1995). The neurological development of prehension: A developmental neurologist's view. *International journal of psychophysiology, 19*, 115–127.

Triesman, A. (1985). Preattentive processing in vision. *Computer Vision, Graphics and Image Processing, 31*, 156–177.

Trevarthen, C. (1984). How control of movement develops. In H. T. A. Whiting (Ed.), *Human motor actions: Bernstein reassessed* (pp. 223–261). Amsterdam: Elsevier.

Trevarthen, C., & Hubley, P. (1978). Secondary intersubjectivity: Confidence, confiding and acts of meaning in the second year. In Lock, A. (Ed.), *1978: Action, gesture and symbol: The emergence of language* (pp. 183–229). New York: Academic Press.

Trevarthen, C., Hubley, P., & Sheeran, L. (1975). Psychological actions in early infancy. *La Recherche, 6*, 447–458.

Uomini, N. T. (2009). The prehistory of handedness: Archeological data and comparative ethology. *Journal of Human Evolution, 57*, 411–419.

Urwin, C. (1982). The contribution of nonvisual communication systems and language to knowing oneself. In Michael Beveridge (Ed.), *Children thinking through language* (pp. 99–128). London: Edward Arnold (Publishers).

Vallbo, A. B., & Johansson, R. S. (1984). Properties of cutaneous mechanoreceptors in the human hand related to touch sensation. *Human Neurobiology, 3*, 3–14.

Vallortigara, G & Bisazza, A. (2002). How ancient is brain lateralization? In L. S. Rogers, & R. J. Andrew (Ed.), *Comparative vertebrate lateralization* (pp. 9–69). New York: Cambridge University Press.

Vallortigara, G., & Regolin, L. (2002). Facing an obstacle: Lateralizatio of object and spatial cognition. In L. J. Rogers, & R. J. Andrew (Eds.), *Comparative vertebrate lateralization* (pp. 383–444). New York: Cambridge University Press.

Vallortigara, G., Cozzutti, C., Tommasi, L., & Rogers, L. J. (2001). How birds use their eyes: Opposite left-right specialization for the lateral and frontal visual hemifield in the domestic chick. *Current Biology, 11*, 29–33.

Vallortigara, G., Rogers, L. J., & Bisazza, A. (1999). Possible evolutionary origin of cognitive brain lateralization. *Brain Research Reviews, 30*, 164–175.

Vallortigara, G., Regolin, L., & Pagni, P. (1999). Detour behavior imprinting, and visual lateralization in the domestic chick. *Cognitive Brain Research, 7*, 307–320.

Vallortigara, G., Rogers, L. J., Bisazza, A., Lippolis, G., & Robins, A. (1998). Complementary right and left hemifield use for predatory and agnostic behavior in toads. *Neuroreport, 9*, 3341–3344.

Vallortigara, G., & Andrew, R. J. (1991). Lateralization of response by chicks to change in a model partner. *Animal Behaviour, 41*, 187–194.

Vandermeer, A. L. H., Vanderweel, F. R., & Lee, D. N. (1995). The functional significance of arm movements in neonates. *Science, 267*(5198), 693–695.

Varney, N. R., & Benton, A. L. (1975). Tactile perception of direction in relation to handedness and familial handedness. *Neuropsychologia, 13*, 449–454.

Ventre-Dominey, J., Nighoghossian, N., & Denise, P. 2003. Evidence for interacting cortical control of vestibular function and spatial representation in man. *Neuropsychologia, 41*, 1884–1898.

Vermeire, B. A., & Hamilton, C. R. (1998). Inversion effect for faces in split-brain monkeys. *Neuropsychologia, 36*, 1003–1014.

Volcic, R. (2008). *Multiple reference frames in haptic spatial processing.* Doctoral thesis, Helmoltz Institute, University of Utrecht.

Volcic, R., & Kappers, A. L. M. (2008). Allocentric and egocentric reference frames in the processing of three dimensional haptic space. *Experimental Brain Research, 188*, 199–213.

Volcic, R., Kappers, A. L. M., & Koenderink, J. J (2007). Haptic parallelity perception on the frontoparallel plane.: The involvement of reference frames. *Perception and Psychophysics, 69*, 276–286.

Vygotsky, L. S. (1962). *Thought and language.* Eugenia Hanfmann and Gertrude Varkar (edited and translated.) Cambridge, Massachusetts: MIT Press.

Wada, J. A., & Rasmussen, T. (1960). Intrscarotid injection of sodium amytal for the lateralisation of cerebral speech dominance. *Journal of Neurosurgery, 17*, 266–282.

Wagner, S. M., Nusbaum, H., & Goldin-Meadow, S. (2004). Probing the mental representation of gesture: Is handwaving spatial? *Journal of Memory & Language, 50*, 395–407.

Walk, R. D., & Gibson, E. J. (1961). A comparative and analytic study of visual depth perception. *Psychological Monographs, 75*, 1–44.

Ward, J. P. (1999). Left hand advantage for prey capture in the Galago (Galago mohili). *International Journal of Comparative Psychology, 12*, 173–184.

Ward, J. P., & Hopkins, W. D. (1993). *Primate laterality: Current behavioural evidence of primate asymmetries.* New York: Springer-Verlag.

Ward, J. P., Milliken, G. W., & Stafford, D. K. (1993). Patterns of lateralized behavior in prosimians. In J. P. Ward, & W. D. Hopkins (Eds.), *Primate laterality: Current behavioural evidence of primate asymmetries* (pp. 43–74). New York: Springer-Verlag.

Warren, D. H., & Pick, H. B. (1970). Intermodality relations in localization in blind and sighted people. *Perception and Psychophysics, 8*, 430–432.

Warren, D. H. (1977). *Blindness and early childhood development*. NewYork: American Foundation for the Blind.

Warrington, E. K. (1982). Neuropsychological studies of object recognition. *Philosophical Transactions of the Royal Society of London, B298*, 15–33.

Warrington, E. K., & Pratt, R. T. C. (1973). Language laterality in left handers assessed by unilateral ECT. *Neuropsychologia, 11*, 423–428.

Warrington. E. K., & James, M. (1967). Tachistoscopic number estimation in patients with unilateral cerebral lesions. *Journal of Neurology, Neurosurgery and Psychiatry, 30*, 468–474.

Washburn, S. L. (1960). Tools and human evolution. *Scientific American, 203*(3), 63–75.

Wassersug, R. J., Naitoh, T., & Yamashita, M. (1999). Turning bias in tadpoles. *Journal of Herpetology, 33*, 543–548.

Watkins, M. J., & Watkins, O. C. (1974). The tactile suffix effect. *Memory and Cognition, 2*, 176–180.

Waters, N. S., & Denenberg, V. H. (1994). Analysis of two measures of paw preference in a large population of inbred mice. *Behavioural Brain Research, 63*, 195–204.

Weber, E. H. (1978). *The sense of touch*. (De Tactu (1834) translated by H. E. Ross, & Der Tastsinn und das Gemeingefuhl (1846) translated by D. J. Murray, for the Experimental Psychology Society) London: Academic Press.

Weinstein, S. (1968). Intensive and extensive aspects of tactile sensitivity as a function of body part, sex and laterality. In D. R. Kenshalo (Ed.), *The skin senses* (pp. 195–222). Illions: Springfield.

Weinstein, S. (1963). The relationship of laterality and cutaneous area to breast sensitivity in sinistrals and dextrals. *American Journal of Psychology, 76*, 475–479.

Weinstein, S. (1962). Tactile sensitivity of the phalanges. *Perceptual and Motor Skills, 14*, 351–354.

Weinstein, S., & Sersen, E. A. (1961). Tactual sensitivity as a function of handedness and laterality. *Journal of Comparative and Physiological Psychology, 54*, 665–669.

Weiskrantz, L. (1989). Blindsight. In F. Boller, & J. Grafman (Eds.), *Handbook of neuropsychology*, Vol. 2 (pp. 375–385). Amsterdam, The Netherlands: Elsevier.

Weiskrantz, L. (1986). *Blindsight. A case study and implications*. Oxford University Press, Oxford.

Weiskrantz, L., Warrington, E. K., Sanders, M. D., & Marshall, J. (1974). Visual capacity in hemianopic field following a restricted occipital ablation. *Brain, 97*, 709–728.

Westling, G., & Johansson, R. S. (1987). Responses in glabrous skin mechanoreceptors during precision grip in humans. *Experimental Brain Research, 66*, 128–140.

Westling, G., & Johansson, R. S. (1984). Factors influencing the finger control during precision grip. *Experimental Brain Research, 53*, 277–284.

Westergaard, G. C., Byrne, G., & Suomi, S. J. (1998). Early lateral bias in tufted capuchins (Cebus paella). *Developmental Psychobiology, 32*, 45–50.

Werner, H. (1958). *Comparative psychology of mental development*. New York: International University Press.

Werner, H. (1957). The concept of development from a comparative and organismic point of view. In D. B. Harris (Ed.), *The concept of development: An issue in the study of human behaviour* (pp. 125–148). Minneapolis: University of Minnesota Press.

White, B. L. (1971). *Human infants*. Englewood Cliffs, New Jersey: Prentice Hall.

White, B. L. (1970). Experience and the development of motor mechanisms in infancy. In K. J. Connolly (Ed.), *Mechanisms of motor skill development* (pp. 95–136). London. Academic Press.

White, B. L., & Held, R. (1966). Plasiticity of sensory motor development in the human infant. In J. Rosenblith, & W. Allinsmith (Eds.), *The causes of behaviour readings in child development and educational psychology* (pp. 60–70). Boston: Allyn & Bacon.

White, B. L., Castle, P., & Held, R. (1964). Observations on the development of visually directed reaching. *Child Development*, *35*, 349–364.

Wills, D. (1979). Early speech development in blind children. *In Psychoanalytic study of the child*, *34*, 85–120.

Williams, H. (1990). Bird song. In R. P. Kesner, & D. S. Olton (Eds.), *Neurobiology of comparative cognition* (pp. 77–126). Hillsdale, NJ: Erlbaum.

Wilson, F. R. (1999). The *Hand. How its use shapes the brain, language and culture*. New York: Vintage Books.

Wilson, M. (2001). The case for sensorimotor coding in working memory. *Psychonomic Bulletin & Review*, *8*, 44–57.

Wishart, U. G., Bower, T. G. R., & Dunkeld, J. (1978). Reaching in the dark. *Perception*, *7*, 507–512.

Witleson, S. F. (1989). Hand and sex differences in the isthimus and genu of the human corpus callosum. *A postmortem morphological study. Brain*, *112*, 799–835.

Witleson, S. F. (1976). Sex and single hemisphere: Specialization of the right hemisphere for spatial processing. *Science*, *193*, 425–427.

Witelson, S. F. (1974). Hemisphere specialization for linguistic and nonlinguistic tactual perception using a dichotomous stimulation technique. *Cortex*, *10*, 3–17.

Witleson, S. F., Glezer, L. L., & Kigar, D. L. (1995). Women have greater density of neurons in posterior temporal cortex. *Journal of Neuroscience*, *15*, 3418–3428.

Wolff, P. (1996). Developmental studies of blind children II. *New Outlook for the Blind*, *60*, 179–182.

Wolley, J. D., & Wellman, H. M. (1992). Children's conception of dreams. *Child Development*, *7*, 365–380.

Wood, H. (1970). *Problems in the development and home care of pre-school blind children*. Unpublished Ph.D. thesis, University of Nottingham.

Woods, A. T., O'Modhrain, S., & Newell, F. N. (2004). The effect of temporal delay and spatial differences on cross-modal object recognition. *Cognitive, Affective and Behavioural Neuroscience*, *4*, 260–269.

Woodworth, R. S. (1898). The accuracy of voluntary movement. *Psychological Review Monograph supplements*, *3*.

Wolpert, D. M., Goodbody, S. J., & Husain, M., 1998. Maintaining internal representations: The role of the human superior parietal lobe. *Nature and Neuroscience*, *1*(6), 529–533.

Yin, R. K. (1969). Looking at upside-down faces. *Journal of Experimental Psychology*, *81*, 351–359.

Yonas, A., & Granrud, C. E. (1985). Development of visual space perception in young infants. In J. Mehler, & R. Fox (Eds.), *Neonate cognition: Beyond the blooming, buzzing confusion* (pp. 45–67). Hillsdale, NJ: Erlbaum.

Young, A. W., & Ellis, A. W. (1979). Perception of numerical stimuli felt by fingers of the left and right hands. *Quarterly Journal of Experimental Psychology*, *31*, 263–272.

Young, G., Segalowitz, S. J., Carter, C. M., & Trehub, S. E. (1983). (Eds.). *Manual specialization and the developing brain*. New York: Academic Press.

Zaidel, E. (1978). Auditory language comprehension in the right hemisphere following cerebral commisurrotomy and hemispheteromy: A comparison with child language and aphasia. In A. Caramazzo, & E. B. Zurif (Eds.), *Language acquisition and language breakdown* (pp. 229–275). Baltimore, Md: John Hopkins University Press.

Zaidel, E. (1977). Unilateral auditry language comprehension on the Token test following cerebral commissurotomy and hemispherectomy. *Neuropsychologia, 15*, 1–18.

Zaidel, E. (1976). Auditory vocabulary of the right hemisphere following brain bisection or hemidecortication. *Cortex, 12*, 191–211.

Zangwill, O. L. (1967). Speech and the minor hemisphere. *Acta Neurologica et Psychiatrica Belgica, 67*, 1013–1020.

Zappia, J. V., & Rogers, L. J. (1983). Light experience during development affects asymmetry of fore-brain function in chickens. *Developmental Brain Research, 11*, 93–106.

Zuidhoek, S., Kappers, A. M. L., Van der Lubbe, R. H. J., & Postma, A. (2003). Delay improves performance on a haptic spatial matching task. *Experimental Brain Research, 149*, 320–330.

Zurif, E. B., & Bryden, M. P. (1969). Familial handedness and left-right differences in auditory and visual perception. *Neuropsychologia, 7*, 179–187.

Zwaan, R. A., Madden, C. J., Yaxley, R. H., & Aveyard, M. E. (2004). Moving words: Dynamic representations in language comprehension. *Cognitive Science, 28*, 611–619.

Author index

Subject index

Most entries are related to the hand. The term 'hand' is therefore not included in the entries.